Great Day Trips to Discover the Geology of Connecticut

by Greg McHone, Ph.D.

Road to Discovery Guides is an imprint of Perry Heights Press.

LCCN: 2004092526

ISBN 0-9630181-4-0

10 9 8 7 6 5 4 3 2 1

Contents

Introduction 1

Photo: Greg McHone

Geologists look at rocks as if each one has its own story to tell. They hope that put together the stories found in rocks will complete a history of the earth—one that will help us understand how the world came to be as we see it today. Some stories are more interesting than others and few can rival the rocks of Connecticut for the stories they tell of how our local region came to be.

All it takes is a short drive in almost any direction to see that the geologic history of Connecticut is long and full of twists and turns. Along nearly every highway are outcrops and road cuts with distinct and different rocks of all kinds. Unique varieties of rock can be found in practically every part of the state—each with its own contribution to make to our knowledge of Connecticut geology.

The Connecticut story

Look at Connecticut's rocks together, and you have a geological story with all the ingredients of a true epic. In their forms, shapes, and colors, is evidence of colossal forces, locked for billions of years in a global tug-of-war, which have made and remade the face of the earth over the vast expanses of deep time.

The oldest rocks reveal continental collisions powerful enough to raise mountains thousands of feet high. Other rocks testify to how the slow, yet relentless effects of weather and the elements eventually all but carried these peaks away. Great columns of stone mark times when the earth cracked and split to spew floods of lava through the crust and over the surface. Sandy beaches and boulder fields along the southeast coast are leftovers of a time in the recent past when Connecticut was frozen beneath immense sheets of ice.

Discovering geology

Connecticut's diverse geology and small size make it an extraordinary natural laboratory. You might drive for days in other states and not see so great a variety or range of geologic features as can be discovered in Connecticut in an afternoon. From Salisbury to Stonington, it is easy to travel over billions of years of geologic time and discover many of the most significant events in earth history.

As much can be learned about geology by traveling around Connecticut as a college student may learn during a semester of lectures—and perhaps more. There are many fine examples of metamorphic rocks in the northwest hills, rocks that were formed under the immense pressures built up during continental collisions. Among them is Stockbridge Marble, a rock formed from lime-rich mud that collected at the bottom of an ancient, extinct ocean.

Sedimentary rocks, made from sand and gravel washed from mountains into lakes and streams, define the Connecticut Valley. These layers record the past in thin slices and reveal startling details of the climate, plants and animals of two hundred million years ago.

Igneous rocks, formed from molten volcanic rock that later cooled and hardened, are also common. Volcanic rocks form the great ridges that trace the backbone of the state today and offer some of the most spectacular local hikes. Still more igneous rocks formed deep beneath the earth's surface and are seen today as the massive granites along the southeast coast.

All in one place

The fabric of Connecticut is a quilt pieced together from the many, different regions of bedrock that underlie the state. With a bit of practice, characteristics of these terrains can be spotted outside the car window, along walking trails in state parks, even in the parking lots of some shopping malls.

The beautifully sculpted rocks of Kent Falls reveal aspects of the old metamorphic terranes of the northwest hills. A seemingly ordinary rock outcrop next to the ball field in Deep River shows a seam where parts of continents that preceded North America and Africa were once pressed together. The Hanging Hills of Meriden stand above the central rift valley that divides the state.

Rocks found in Connecticut have all these stories to tell and more. It is amazing how much there is to discover about our earth history in the state—if you know where to look.

The history of science

As it turns out, there is more to the story of the geology of Connecticut than just natural history. The human stories that go along with the science are equally epic. America's first generation of geologists began their training in Connecticut when Thomas Jefferson was in his first term as President. They pioneered natural science in the New World not long after Meriwether Lewis and William Clark made their expedition west in search of a route to the Pacific. The work they began nearly two centuries ago laid a foundation for the knowledge of Connecticut geology today.

How to use this book

This book was developed to make it easy for anyone to discover and learn about geological science. Connecticut is a small state, but its rocks have been in the middle of many of the most important events in earth history. The state is full of geologic features that are fun to explore and provide excellent examples of the fundamental principles and knowledge of the field.

Viewed in the local context, geology is more accessible, more intriguing—and easily understood. Take advantage of Connecticut's natural laboratory simply by visiting a few of the locations described in this book.

This guide has two parts. The first few chapters tell the story of the geology of Connecticut and give a background about the scientists, discoveries and ideas that frame current knowledge of how the state was formed. The second portion of the book describes field trips to local sites where these fundamental concepts come to life.

To get started making your own explorations, thumb through the day trip descriptions together with your friends or family to find a few that look like good places to start. The trips are geographically dispersed and include sites from parks to malls. Decide on a few that fit your group's interests or plans.

Once a trip has been identified, return to the background chapters at the front of the book. Read the many remarkable stories about how scientists unraveled the mysteries hidden in the earth's crust and how local scientists made important contributions working in Connecticut. It is possible to discover over a billion years of earth history in just a few trips.

What can be learned will not only add to the enjoyment of the trips—but can change the way the landscapes of Connecticut are seen outside the car window.

History & Geology 2

The Scottish Geologist James Hutton
© The Natural History Museum, London

G eology measures the vastness of deep time, but it is actually a relatively new field of science. While humans have explored nature and our surroundings for hundreds of thousands of years, much of what we know of the world has been learned in the past few centuries. Essential pieces of the puzzle have fallen into place in the past few decades—and more is still unknown.

Geologic time

The great challenge of geology has always been to gain a grasp of time—to know not only *what* events have occurred in the history of the earth, but *when* they occurred. Before scientists formed a good estimate of the age of the earth and the pace of change since the planet first coalesced, much of what was learned about mountains and valleys and the rocks underlying them was of little value.

For many centuries, European philosophers believed that the earth was a relatively young planet, barely 6,000 years old. Early estimates of the age of the earth were influenced by biblical accounts of the Creation and by the work of an Irish bishop, James Ussher, who used biblical references to calculate that the first day of the Creation had been October 24, 4004 B.C.

It was not until after the turn of the eighteenth century that concepts of deep time—and an ancient earth—took shape. A Scottish gentleman, James Hutton, is credited with having made many significant contributions to the modern knowledge of geology and to efforts to know the age of the earth with the publication of his *Theory of the Earth* in 1795.

Efforts to understand the progress of time on earth, or geologic time, showed the problem had two components. There was relative time, the age of one rock or group of rocks judged against others nearby, and there was absolute time, the age of rocks measured against a standard such as years. Relative time could be judged more easily than absolute time. Even before the age of the earth was well understood it was possible to look at rocks that occurred together and to recognize which were older and which were younger.

Relative time

The Danish philosopher Nicholas Steno made early explorations of the concept of relative time in the late-1600s. His interest had been piqued by studies he made of mysterious "tongue stones," curiosities that later came to be recognized as fossilized shark teeth. Studies Steno made of mountains in Italy where tongue stones were found led him to propose a series of ideas that formed a basis for much of geology.

Looking at the layers of rock, or strata, that he encountered in the mountains, Steno realized that individual layers marked different periods of geologic time. Lower layers,

Layers of rock, like ones pictured above from the Powder Hill Dinosaur Park in Middlefield, are known as strata. Photo: Greg McHone

near the bottom of a set of strata, had accumulated first and were relatively older. Upper layers had to have been added later and thus were relatively younger. This concept is known today as *the principle of superposition of strata*. By proposing this law, along with other fundamental principles, Steno worked out a way to judge not only the relative age of rock strata, but also to recognize sequences of events that occurred since they formed.

Hutton began making similar explorations a century or so later after noticing layers of rock that occurred around his home in Scotland. It was clear to him that strata found in the region were formed from mud and sand, which later came to be naturally cemented into solid rock.

Hutton recognized also where younger layers of rock had later cut across this ancient surface and how over time the strata had been become tilted and eroded. It was apparent to Hutton that great lengths of time would be required in order for natural processes to form layers of mudstone and sandstone and for the strata to later be tilted or eroded. The mud and sand first had to be turned to stone. The landscape must then have somehow been raised or lowered for the sandstone to become tilted and fractured. Finally, fragments must have remained exposed on the surface long enough to be worn away by weather and the elements.

This "successions of former worlds," as Hutton described geologic events, required more than a few thousand years. The fact that these rocks existed led Hutton to conclude the earth must be much older than previously believed. Hutton had no idea of just how old the earth might be, but he was convinced it was ancient. "We find no sign of a beginning," Hutton wrote, "and no prospect of an end."

Hutton concluded that "the present is the key to the past," that geological processes that worked to form and deform rocks in the past must be the same as those at work in the present, and that the pace of geologic change must therefore be immensely slow.

He chose the term "Uniformitarianism" to describe his view of a world that was slowly, but continually changing. The volcano known today as Mount Vesuvius provides an example of Hutton's perspective. Vesuvius has erupted several times over the past 2,000 years of recorded history. Yet, over what seems to be a long period of time, its dimensions have hardly changed. Based on its observed rate of activity, eruptions would have to continue for hundreds of thousands of years for the volcano to reach its current size.

Absolute time

After Hutton proposed his theory of a far more ancient earth, the challenge for geologists was to work out an accurate estimate for just how old the planet earth truly was. William Smith advanced the search for a measure of absolute time in the early-1800s. Smith was an engineer by trade and an amateur fossil hunter by avocation. His work digging canals allowed him to indulge both interests and so Smith came to know the layers of bedrock underlying much of the British Isles as well as fossils associated with them. The oldest rocks contained no evidence of fossils. Middle-aged rocks contained fossils of sea life, while still younger strata contained fossils of life on land.

Smith published his findings in a map of the geology of the British Isles in 1815. For the first time, his map organized rocks in a chronological order according to the fossils associated with them. The principle of the superposition of strata made it possible to judge the relative age of rocks found within connected areas. Smith's map created a system for ordering rocks according to a regional and eventually a global timescale.

Smith made use of the fact that fossils, unlike rocks, are limited in their occurrence to discrete periods of time. The fact that various animals appeared at certain times in the past, and later disappeared or went extinct, is true everywhere. Wherever fossils were found they provided a natural order that could be used to place rocks in time. "Once dinosaurs die out, they can never return again," paleontologist Stephen Jay Gould once pointed out. "Whereas quartz can be formed at any time…"

The geologic column

Smith's map plotted the first section of what became a timescale for the ages of the earth. His work led to geologic time being plotted on a scale of years, known today as the geologic column, and to a new field that correlated rocks with fossils known as biostratigraphy.

The work Smith began is carried on today by geologists and paleontologists who study the past using fossil evidence. International commissions have since been convened to integrate new findings and set standards for an increasingly detailed global chronostratigraphic timescale. These timescales divide the past into many different periods. Included among these are the Paleozoic Era (the time of primitive life), the Mesozoic Era (the time of middle life), and the Cenozoic Era (recent time).

Physical evidence for many occurrences and periods of time

is often missing or incomplete—in some cases due to the effects of the same geological processes and changes we seek to understand. "The deeper we go into geological prehistory," Yale geologist John Rodgers once wrote, "the more difficult it becomes to decipher the record of past events."

The development of natural science

Hutton published his theory and Smith produced his map during a time when many significant advances were beginning to be made in fields of natural science. The scholarly study of nature gained popularity among all sorts of people in the early-1800s, from painters to poets and from amateur enthusiasts to brilliant scholars. This new interest led to important advances in geology and in fields such as paleontology (the study of the past from fossils) and biology (the study of life).

New approaches to classifying and naming plants and animals increased the knowledge of living things, while observations made of rocks and fossils led to a new understanding of the diversity of life and of the expanse of geologic time. In the 1750s, for example, the Swedish philosopher Carolus Linnaeus developed new systems for studying and classifying living things and made great progress in describing the diversity of life on earth. In the 1790s, the French scientist Georges Cuvier made studies of fossils that revealed previously unknown forms of life that had existed in the past and long since gone extinct.

Science in the New World

The discoveries made by Hutton and Smith helped to usher in the "Heroic Age" of geology, as the period between 1790 and 1830 is dubbed. European scientists made great strides toward understanding the age and history of the earth during this time.

By comparison, science in the New World was in its infancy. The United States was a new and young nation, preoccupied with the business of establishing its democracy. The Revolutionary War was still fresh in the minds of many Americans. The nation had a long way to go to match the power of other nations—and to catch up with scientific advances being made in Europe.

American medical schools taught sciences like chemistry and anatomy to aspiring physicians, but a quarter century after the Revolution no college offered classes in geology. America's first geologists and paleontologists were self-taught and reliant on their wits and intuitions.

Benjamin Silliman
Courtesy of the Peabody Museum of Natural History, Yale University, New Haven, CT

Geology in Connecticut

An important development came in 1802, when Yale College in New Haven hired a young graduate, Benjamin Silliman, to begin teaching a class in chemistry and natural history. Within a few years, Silliman made Yale the first American college to offer geology and he established a mineral and fossil collection that would come to include the first verifiable dinosaur bones found in North America.

In the fifty years that followed, Silliman went on to pioneer the sciences of geology, mineralogy and paleontology in the United States. He became a leading advocate of science in early nineteenth century America, and inspired many people, from ordinary citizens to scholars, to become interested in various fields.

The geologists he trained came to have a profound influence on the geological sciences in New England and around the world. Studies they made of Connecticut and neighboring states contributed greatly to the knowledge of the geology of the region.

Yale University Art Gallery, Gift of Bartlett Arkell, B.A. 1886, M.A. 1898, to Silliman College

Benjamin Silliman

Benjamin Silliman devoted his life to seeing that people from all walks of life took a bit of science away with them. From the Yale students that packed his classes, to crowds that filled public halls to hear him lecture, he shared his passion for science and the study of the natural world with everyone he could.

Before Silliman, Yale was a college dedicated to the study of the classics. By the time Silliman retired fifty years later, Yale had come to be highly regarded for its science department and the many great scientists educated there. "Now that I have traveled from Niagara to Georgia," the renowned English geologist Charles Lyell wrote to Silliman in 1842, "and have met a great number of your countrymen. . . I may congratulate you, for I never heard as many... refer as often to any one individual teacher as having given direction to their taste."

As the Heroic Age of geology gathered steam across the Atlantic in England, Silliman trained the first generations of American geologists in Connecticut. The scientific journal he founded, *The American Journal of Science*, opened a dialogue between the English and the Americans and advanced the progress of geological science in the United States.

Silliman wrote the first scientific study of Connecticut geology, made the first studies of the space rock now famous as the "Weston meteorite" and described fossil fish he found in the valley.

In writing about the geology of New Haven, Silliman came to recognize the differences between the ancient metamorphic rocks of the highlands to the east and the west and the much younger, sedimentary rocks and volcanic, "traprock" ridges of the valley. He recognized similarities between the geologic features of volcanic ridges he'd seen in Scotland and local volcanic rocks such as East Rock in New Haven.

Perhaps the best measure of his accomplishments was his influence on his students. His son Benjamin Jr., and son-in-law James Dwight Dana also became professors at Yale and carried on the work he started. Edward Hitchcock of Massachusetts, who came to New Haven to study with Silliman, later completed the first geological survey of Massachusetts and became president of Amherst College. Hitchcock also made the first studies of the famous dinosaur footprints commonly found in the Connecticut Valley.

James Gates Percival and Charles Upham Shepard made the first systematic survey of the geology and mineralogy of Connecticut. Their work remains startlingly accurate. Amos Eaton was professor of geology at Williams College and Rensselaer Polytechnic Institute and conducted the geological survey of New York State. Denison Olmstead supervised the geologic survey of North Carolina.

Edward Hitchcock

Edward Hitchcock battled illness throughout most of his life to become one of the greatest geologists and paleontologists ever to study the Connecticut Valley. As a young man, he collected rocks, fossils and plants and taught himself to be an expert on natural history. Despite his lack of formal scientific education his observations of the geology of the north valley earned him notice.

Hitchcock was driven always by a great hunger for learning, but the more he sought to satisfy his curiosity—for religion, for classic literature, for ancient language and for science—the worse the pains in his belly grew to be.

"I had acquired a strong relish for scientific pursuits," he recalled of his youth, "and I seized upon every moment I could secure—especially rainy days and evenings—for those studies. I was treated very leniently by my father and brother, who probably did not know what to do with me, but saw plainly that I should not become distinguished as a farmer."

At the same time, Hitchcock was plagued by stomach ailments. In 1814, a bout of the mumps left him blind and forced Hitchcock to abandon his dream of a college education.

He regained his sight in 1816 and was taken in by Deerfield Academy where he resumed his studies. "I laboured intensely," he wrote, "to maintain myself in spite of a defective education, weak eyes, and poor health."

Hiking in the mountains around Deerfield proved more therapeutic than any doctor's treatment, and the outings developed in him a fascination for natural history that occupied him the rest of his life. He made frequent field trips to learn all he could about the natural history of the valley. He began attending lectures at Amherst College and became acquainted with many local scientists, including Amos Eaton, who had been a student of Silliman's at Yale College.

Through his observations and writings he opened a correspondence with Benjamin Silliman. In 1818, Hitchcock became one of the earliest contributors to the *American Journal of Science* when Silliman published his report on the geology and mineralogy of the north valley. He wrote prolifically and his reputation quickly grew.

After Hitchcock was offered the position of professor of Chemistry and Natural History by Amherst College in 1825, he traveled to New Haven to spend several months studying with Silliman. He returned to Amherst to begin a distinguished career as professor of natural history and geology and was appointed Geologist to the State of Massachusetts. Hitchcock later served as president of the college and made important contributions to the development of the science department at that institution.

Among his observations, Hitchcock recognized differences in the traprock ridges prominent up and down the valley. He proposed that some were formed by lava pressing up from cracks, or faults, in the earth's crust, rather than exploding from a volcano. It took years for his ideas to be accepted, but Hitchcock had anticipated what geologists know today as fissure eruptions.

Hitchcock is perhaps best known today for his studies of the many dinosaur footprints commonly preserved in the sedimentary rocks of the Connecticut Valley. In his understanding of the valley's ancient past, Hitchcock was without peers. He deduced from its sediments much of what paleontologists have since confirmed to be true. "The probability is that the climate, during the sandstone period, was tropical," he wrote, "with perhaps an alternation of wet and dry seasons."

Hitchcock realized also that the valley once held large bodies of water and described how these lakes supported large and diverse communities of land-dwelling or terrestrial animals. He believed many of the animals that left tracks in the valley were large, flightless birds. Paleontologists later recognized similarities in the designs of birds' feet and certain types of dinosaurs and identified many of the footprints as those of dinosaurs.

The footprint Hitchcock described as Eubrontes, made by a dinosaur
Courtesy, Paul E. Olsen

The first survey

The Connecticut state legislature authorized the first geological survey of the state in June 1835. James Gates Percival was charged with mapping the geology of the state and Charles Upham Shepard given responsibility for reporting on local minerals. It took Percival over seven years to publish his map, an incredible accomplishment given that he surveyed the entire state. The map he produced recognized all the major geologic components of the state and remains remarkably accurate.

James Gates Percival

The Geological and Natural History Survey of CT

Few people realize it, but Connecticut has been a leading state for environmental science for more than a century. The legislature established the State Geological and Natural History Survey of Connecticut in June 1903 and funded fundamental geological and biological research for many decades. Scientists from the state's leading educational institutions, including Connecticut Agricultural College (now the University of Connecticut), Trinity College, Yale University and Wesleyan University, cooperated on the research and shared the survey's mission:

"First, the advancement of our knowledge of the geology, botany, and zoology of the state as a matter of pure science; second, the acquisition and publication of such knowledge of the resources and products of the state as will serve its industrial and economic interest; third, the presentation of the results of investigation in such form as to be useful in the educational work carried on in the various schools of the state. These three aims, the purely scientific, the economic, and the educational, we have endeavored to keep in mind in all plans which have been made."

Hundreds of scientific studies were published by the survey, many of which are still available today. A part of the state's Department of Environmental Protection since 1971, survey publications are available at the DEP Store at the Department's offices in Hartford.

The survey's superintendents & directors

According to Robert J. Altamura, who wrote a history of the state survey in 1989, one of the first tasks undertaken by the survey was to collect a variety of scientific publications, information published by local universities, and to assemble the work into the Manual of the Geology of Connecticut. The task was led by William North Rice, who served as the survey's first superintendent. Rice had also been the first to receive a Ph.D. in geology from an American educational institution, with the degree Yale awarded to him in 1867.

Rice went on to become professor at Wesleyan and along with his colleagues carried on the tradition of geological research Silliman had begun decades earlier. He was succeeded over the years by many of the most accomplished geologists in the country. The survey, under the leadership of another Wesleyan geologist, Joe Webb Peoples, played a vital role in the conservation of the dinosaur footprints discovered in Rocky Hill Connecticut in 1966 and in the establishment the Dinosaur State Park months later. Today, the survey continues to provide essential environmental information to individuals and communities working to conserve and protect natural resources and environments around the state.

The Geologic Map of Connecticut

One of the great geologists of recent generations was John Rodgers, a professor emeritus at Yale University. Rodgers liked to say that he studied "the comparative anatomy of mountain ranges," but over the course of his life he did much more.

Rodgers traveled the world, from Connecticut to Russia, to study geology and mountains. He was an expert on stratigraphy, the field of geology that deals with layers of rock like those found in the Connecticut Valley.

Ideas Rodgers helped to develop about the mountain-building events of Connecticut's deep past have since become the basis for the modern view of the state's geological history. The map he published of the bedrock geology of Connecticut in 1985 is the most complete map of the state ever produced. *The Generalized Bedrock Geologic Map of Connecticut* is widely used today and is an essential guide for any geologist, from scholars to school kids.

Rodgers also served as editor of the journal Silliman had started, *The American Journal of Science*. He was also a musician and recorded a jazz composition based on the "music of the planets."

John Rodgers
Copyright © 1977, Peabody Museum of Natural History, Yale University, New Haven, CT

The Geology of Connecticut 3

Photo: Greg McHone

"There is a connection between people and land and it is just as important today and tomorrow as it was in the past. Although the connection has changed, our need to understand more about it has not."
—Michael Bell, *The Face of Connecticut*, 1985

Our three-dimensional world moves steadily forward along a fourth dimension we know as time. As long as there is heat energy in the universe, atoms will vibrate, clock hands will rotate and time will go on. After it coalesced from the ashes of still older systems, the planet has traveled over a vast span of time, some 4.6 billion years. The earth's history goes far beyond human history and to the limits of our comprehension.

In Connecticut, local landscapes and bedrock record large portions of earth time, from well over a billion years ago to the present. Near Danbury, I-84 crosses some of the most ancient rocks in the state. Portions of the North American continent have been in place here for 1,300,000,000 years—so long ago the zeroes now only obscure their age. The brownstones common along roadsides in the Connecticut Valley were deposited over 220 million years ago, around the time dinosaurs appeared. Most of the farmland, marshes, lakes, and streams we see and enjoy around the state today are much more recent features, formed only in the last 16,000 years or so, since the end of the last Ice Age.

The story Connecticut rocks tell is one of gradual changes that occurred over great lengths of time. Its involves ideas like those that scholars such as Steno and Hutton first explored centuries ago, but it is also a story that until recently was missing essential parts.

Time & change

Among Hutton's many contributions to geology was a model of geological change that has come to be known ever since as "the rock cycle." It is a representation of how the earth's crust is constantly recycled and helps to explain how many rocks, including varieties commonly found in Connecticut, could be formed.

New rock is formed volcanically, as hot, molten matter presses up from the earth's mantle to places on or near the surface where it cools and hardens. Volcanic and igneous rocks are common in Connecticut; the most common forms include the rocks known as basalt, dolerite, and granite.

Rock that is exposed at the surface is slowly worn down by weather and the elements. Sedimentary rocks are formed from mineral sediment, sand or gravel worn from older rocks, and from organic material, such as seashells and plant mold. These sediments may accumulate in layers and slowly be turned to stone by processes of natural cementation. Sedimentary rocks are common in the Connecticut Valley and include rocks known as arkose and shale.

Metamorphic rocks form when igneous or sedimentary rocks are subjected to great heat and pressure and changed or recrystallized into new and different forms. Metamorphic rocks

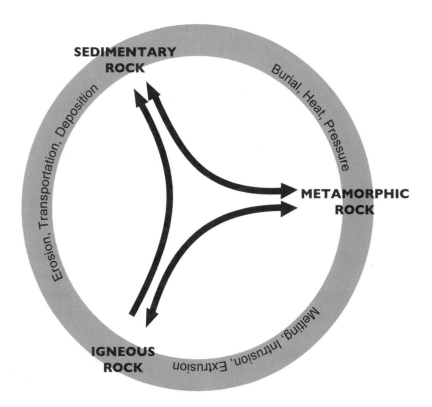

The Rock Cycle

are the most common in Connecticut and include forms such as marble, gneiss and schist.

Continental drift

Perhaps the most important development in geology since Hutton's theory of an immensely old earth, began as an equally radical idea known as "continental drift." For centuries, people from geologists to schoolchildren have looked at maps of the world and noticed how the continents seem as if they could fit together like pieces in a jigsaw puzzle. The shapes of the western shorelines of Africa and Europe, for example, fit the shapes of the east coasts of North and South America as if they were made for each other.

These similarities have seemed to many to be too great to simply be coincidence.

Among those who were intrigued by the similarities of the shapes of the continents was renowned meteorologist Alfred Wegener. Early in the twentieth century, Wegener began making studies of these peculiar quirks of nature. He collected a great deal of information about the way the geological features of continental puzzle pieces matched up. What Wegener discovered led him, along with several of his more bold colleagues, to propose the world had once been united as a great "super-continent" that later split in several pieces to form the continents as they are seen today.

The idea came to be known as the theory of continental drift. As has often been the case when radical, new ideas are proposed, many scientists dismissed the notion that landmasses as large as continents could be moving as outlandish and bizarre. Worst of all, they argued, Wegener failed to account for a force or forces

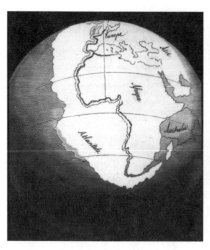

Before and after: In 1858 Antonio Snider-Pellegrini made these maps of how the American and African continents may once have fit together.
Reproductions of the original maps courtesy of University of California, Berkeley.

great enough to cause continents to move as he claimed they had. The theory was left to languish until decades later, when explorations of the oceans returned unexpected results.

After the end of the Second World War, scientists began focusing their attention on the oceans and the seafloor beneath. New developments in submarine warfare gave new urgency to the development of systems for detecting submerged submarines, as well as understanding how weather patterns and ocean currents affected naval operations. It soon became necessary to

Alfred Wegener
Courtesy of the Alfred Wegener Institute for Polar
and Marine Research, Bremerhaven, Germany

develop techniques for distinguishing the magnetic effects of the steel hulls of ships and submarines from the natural magnetism of the crust beneath the seafloor. Careful surveys of the crust beneath the world's ocean began to be made and in the 1950s curious features began to emerge.

Hundreds of "stripes" of strong and weak magnetism appeared to be parallel and roughly symmetrical on either side of a great, undersea mountain belt extending down the middle of the Atlantic Ocean. The belt was known to be volcanic from areas where it could be examined, such as Iceland. Samples of ocean crustal rocks with weak magnetism showed they had formed during a time when the magnetic poles of the earth were actually reversed: when the "north pole" of a compass needle would have pointed south. The fact that the earth's magnetic field had reversed many times in the past was by then already known, but the causes for these changes are still mysterious.

Seafloor spreading

Samples of ocean crust were also dated using new methods of radiometric analysis. The results showed that the seafloor was youngest close to the central mid-ocean ridge of undersea mountains. Samples taken further away from the ridge and closer to the continents grew progressively older. By 1960, Dr. Harry Hess of Princeton University, along with several other American geophysicists, proposed that new ocean crust was constantly being formed by volcanism along the center of the mid-ocean ridge. This new crust gradually split up the middle and migrated away from the ridge along great "transform" fractures.

As new volcanic ocean crust cooled, its magnetic properties were influenced by the positions of the earth's magnetic poles. Analyses of these magnetic patterns proved the seafloor was moving. This process is today known as "seafloor spreading" and the evidence for it is irrefutable.

Seafloor spreading works to move more than just newly formed crust; it forces the continents away from mid-ocean ridges as well. One way to visualize this process is to imagine it running in reverse. The continents would close in on each other until they fit together in the way they had

Harry Hess
Courtesy, Department of Geosciences,
Princeton University

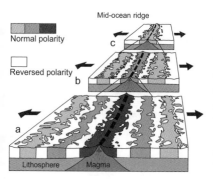

Mid-ocean ridge

Normal polarity

Reversed polarity

Lithosphere   Magma

As new ocean crust cools, its magnetic properties record the positions of the earth's magnetic poles. Studies of these rocks provided evidence that seafloor spreading was a force that contributed to continental drift.
from This Dynamic Earth, United States Geological Survey

before seafloor spreading forced them apart. While the theory of continental drift today seems like common sense, it is unfortunate that Alfred Wegener did not live to see his ideas vindicated.

Plate tectonics

The movements of continents have since been attributed to a process described as "plate tectonics," one that is now accepted as having played a fundamental role in shaping global geology—as well as the construction of eastern North America and Connecticut.

The term plate tectonics is composed of two words. Tectonics simply refers to earth movements, whether large or small, slow or fast. An earthquake, for example, is a tectonic event. Faults, or the cracks in the earth's crust along which earthquakes occur, are tectonic features. Uplifts that work to build mountains are also tectonic events and may be accompanied by folding of the earth's crust referred to as tectonic folds.

Plate tectonics are in a category of earth movements all their own. The word "plates" refers to large sections of crust and uppermost sections of mantle that move along with them. The movements of these packages of crust and mantle plates explain how oceans and continents were made and remade over time.

But for continents and new ocean crusts to be in motion, there

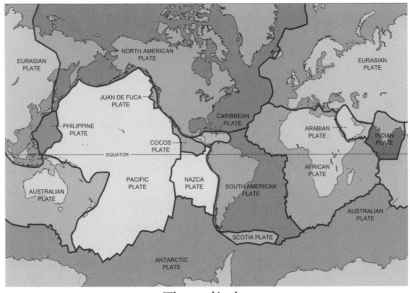

The earth's plates
from This Dynamic Earth, United States Geological Survey

had to be more going on than just seafloor spreading. The earth has stayed about the same size since it formed four and a half billion years ago. To make room for new ocean crust pouring from mid-ocean ridges, old crust had to be destroyed somewhere else.

A new way of thinking

The advent of plate tectonics forced geologists to completely reevaluate the way they thought about geologic processes. We know now that the destruction of old crust occurs primarily around the margins of oceans, near continents, next to deep trenches and island arcs. As relatively thinner, newly created ocean crust moves away from mid-ocean ridges, it eventually comes to push up against older, relatively thicker, continental crust or ocean crust. Where this occurs, the section of ocean crust that is in motion descends along boundary lines, sliding under the adjacent crust and down into the mantle of the earth. Deep ocean trenches are formed, caused by the drag of downward-sliding crust. Curving chains of islands and continental volcanoes, such as the famous "Ring of Fire" surrounding the Pacific Ocean, may form above deep areas of crustal destruction.

The collision boundaries on opposite sides of plates are called subduction ("moved underneath") zones. Examples of such zones include the Aleutian Trench and islands west of Alaska (where the Northern Pacific Plate is subducted), and the Puerto Rico Trench and islands of the eastern Caribbean (where part of the western Atlantic is subducted). Iceland and a few other islands emerged from the ocean along the divergent boundary between the eastern and western North Atlantic. Even more significant are mountain chains such as the Appalachians and Alps, which formed where continental plates collide. Unlike ocean plates, the continents are too thick and too light in weight to dive into the mantle. They crumple up along the convergent boundary instead.

This crumpling causes structures and minerals of the crust to be remade by metamorphism and mountain-building events known as orogenies to occur. The famous Himalayan Mountains of India and Tibet, including the world's tallest peak, Mount Everest, are being formed as part of an orogeny that is ongoing today. The

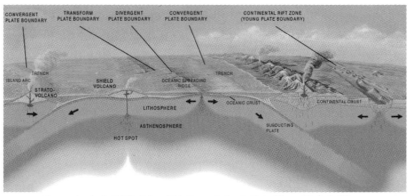

Cross section by José F. Vigil from This Dynamic Earth, *United States Geological Survey*

mountains are being built as the Indian Plate continues to push northward into western Asia. Some mountain chains, including the Appalachians, actually record several orogenies, each caused by an ancient plate collision.

Plate tectonics explain how many volcanoes and earthquakes originate. Where plates diverge and converge, chains of volcanoes erupt from magmas formed beneath the margins. Subduction zone volcanoes can be violent, as represented by Mount Pinatubo in the Philippines, and Mount St. Helens in Washington. Most earthquakes occur along subduction zones and along plate boundaries that slide horizontally by one another, as is occurring in southern California. Earthquakes occur when a fault moves, and active faults essentially define plate boundaries.

Maps of the Ancient Earth

The concept of plate tectonics has since provided geologists with many powerful tools. Since the direction and speed of plate movements have become known it's become possible to work backward and calculate the positions of large landmasses at different times in the past. This work is helped by studies of orogenies in the continental mountain belts formed by plate collisions. Magnetic properties of rocks can be used to discover how far north or south rocks were when formed, or their original latitude. These techniques allow geologists to go back a billion years or more and reconstruct the earth's past. It becomes very difficult to go back further because so much has happened and changed in the years since.

Maps of the earth in the past are very helpful to understanding how New England was assembled. Geologist Ron Blakey, Professor of Geology at Northern Arizona University, first became interested in paleogeography while studying the stratigraphy of the Colorado Plateau. Since then, Ron has created many maps to show how continents have drifted and been reshaped over time.

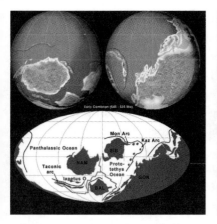

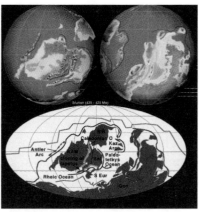

The ancient continents, Proto-North America (NAM), Baltica (BAL), Gondwanaland (GON), Proto-Siberia (SIB), as they were between 545-535 million years ago, in the Early Cambrian Period. Copyright © Ron Blakey.

The continents as they were between 435-425 million years ago, in Silurian time. Proto-North America and Baltica are on course to collide—and to close an ancient ocean, Iapetos, in the process. Copyright © Ron Blakey.

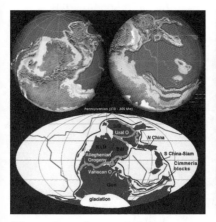

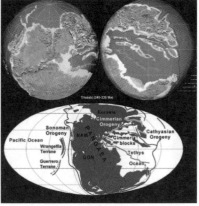

The continents between 310-300 million years ago, in Pennsylvanian time. Gondwanaland is set to collide with Proto-North America and Baltica. Copyright © Ron Blakey.

By Early Triassic time, the continents were pressed together to form the super-continent of Pangaea. Near the center of this vast landmass was the region that would come to be known as Connecticut. Copyright © Ron Blakey.

The terranes of Connecticut

The fabric of Connecticut weaves together a surprisingly diverse group of geological regions, or terranes, each with its own story to tell. The bedrock of the state is composed of several distinct and different terranes, most formed during a series of plate collisions that occurred along the eastern edge of the ancient continent known today as Proto-North America.

Long periods of erosion have since smoothed and flattened Connecticut to where today it appears fairly uniform, but the state includes as many as nine separate terranes, depending on how they are defined. The boundaries between these terranes are often, but not always, marked by major faults along which crustal pieces slid into one another.

The detailed histories of the terranes are still being worked out, but when these areas were first put together on one map in the 1970s, it created a sensation among the close-knit, local community of geologists. A simplified map of the terranes of Connecticut appeared in *The Face of Connecticut* by Michael Bell.

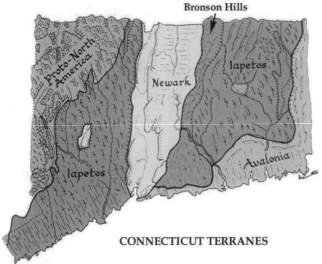

CONNECTICUT TERRANES

Courtesy, Connecticut Department of Environmental Protection, copyright 1985

The Proto-North America Terrane includes the most ancient rocks in the state, similar to those found farther west and north, in western Vermont and the Adirondacks. Large portions of the state, primarily the western and eastern highlands, were formed later, as the seafloor beneath an ancient ocean known as Iapetos was folded into Proto-North America. These areas are known as the Iapetos Terranes. The Bronson Hills Terrane was formed when volcanic island arcs impacted the continent and metamorphosed portions of Iapetos Ocean crust. Avalonia was formed following still another impact, by the collision of a fragment of an ancient, continental volcanic chain.

The Newark Terrane is the youngest of all and has a very different history. Rather than being formed by continental collisions, the valley resulted from continental rifting, or the stretching of continental crust. It is distinguished by the layers of sediments and lava later deposited over Iapetos rocks and is similar to so-called rift basins widespread today across eastern North America and Africa.

The renowned Yale geologist, John Rodgers, showed the results of this long series of events in great detail in his 1985 geologic map of Connecticut. Like pieces of our own local jigsaw puzzle, it shows how the ancient terranes were joined together from west to east.

The rocks of Connecticut

Hutton proposed that many of the earth's features could be understood by observing geologic processes at work in the present. Looking around Connecticut, for example, it is easy to imagine how the sedimentary rocks of the valley were formed. We can see where mud and sand was carried by rivers and streams from higher areas and deposited in layers, or strata, in lower areas. Similarly, organic matter suspended in lakes settles to the bottom as layers of thick mud or clay. Water trapped in these sediments often contains dissolved minerals that penetrate up from deeper strata into

GENERALIZED BEDROCK GEOLOGIC MAP OF CONNECTICUT

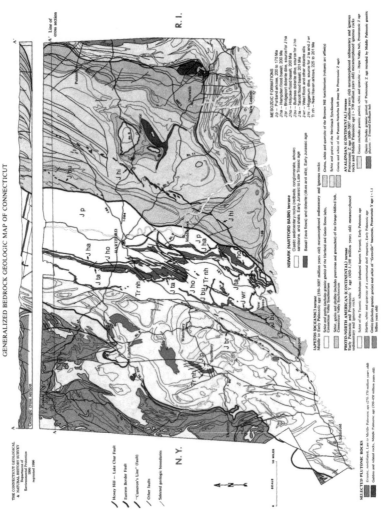

THE CONNECTICUT GEOLOGICAL
& NATURAL HISTORY SURVEY
Department of
Environmental Protection
1985
reprinted 1996

A — Line of cross section

A' — Line of cross section

Geologic cross section

N.Y.

R.I.

N

SCALE
0   10 MILES

Honey Hill – Lake Char Fault
Eastern Border Fault
"Cameron's Line" (fault)
Other fault
Selected geologic boundaries

SELECTED PLUTONIC ROCKS

Granite, metafoliated, Late to Middle Paleozoic age (230-370 million years old)
Gabbro and related rocks, Middle Paleozoic age (380-430 million years old)

MESOZOIC FORMATIONS

Jp — Portland arkose, 200 to 176 Ma
Jha — Hampden flood basalt, 200 Ma
Jbr — Bridgeport dolerite dike, source for Jha
Jho — Holyoke flood basalt, 200 Ma
Jbu — Buttress dolerite dike, source for Jho
Jta — Talcott flood basalt, 201 Ma
Jwt — West Rock and other dolerite sills
Jwr — Kagganum sills, source for Jta and Jwr
Trnh — New Haven arkose, 225 to 201 Ma

IAPETOS (OCEANIC) terrane
Middle to Early Paleozoic age (350–500? million years old) metamorphosed sedimentary and igneous rocks

Schist and gneiss (includes granitic gneiss) of the Hartland and Gneiss Dome belts.

Schist, granite and phyllite (includes gneissic granite and granodiorite) of the Orange-Milford belt, Connecticut Valley Synclinorium

PROTO-NORTH AMERICAN (CONTINENTAL) terrane
Early Paleozoic and Proterozoic "Z" age (450–1,100 million years old) metamorphosed sedimentary and igneous rocks

Schist of the Taconic Allochthon (displaced Iapetos Terrane), Early Paleozoic age

Marble, schist and quartzite of a continental shelf sequence, Early Paleozoic age

Gneiss (includes granitic gneiss) and schist of "Grenville" basement, Proterozoic Y age (~1.1 billion years old)

NEWARK (HARTFORD BASIN) terrane
Clastic sedimentary rocks (redbeds: conglomerate, arkosic sandstone, and shale. Early Jurassic to Late Triassic age

Basalt (lava flows) and dolerite (dikes and sills), Early Jurassic age

AVALONIAN (CONTINENTAL) terrane
Proterozoic "Z" age (570–620 million years old) metamorphosed sedimentary and igneous rocks and Middle Paleozoic age (~ 370 million years old) metamorphosed igneous rocks

Gneiss (includes granitic gneiss, schist and quartzite — Hope Valley belt, Proterozoic Z age

Gneiss (includes granitic gneiss and schist) of Proterozoic Z age intruded by Middle Paleozoic granitic plutons — Putnam-Dudham belt

Courtesy, Connecticut Department of Environmental Protection

more shallow layers, to where they act as natural cements. Sand is cemented into sandstone, mud into mudstone and clays into shale—sedimentary rocks that are abundant in the Connecticut Valley. Similarly, we can look to volcanic events occurring today in places such as Hawaii or Iceland for help imagining how the traprock ridges formed. We can make chemical and physical analyses to see how ancient lava flows, such as the basalt columns preserved in some of Connecticut's traprock ridges, compare with lavas and volcanic ash that flowed and hardened more recently.

Processes that worked to form other local rocks are not as easily observed. A variety of analytical tools must be used instead. Chemical and mineral analyses of granite and related rocks, such as gabbro and diorite, show they are similar to volcanic rocks. The evidence suggests they were formed from magma, or molten rock, that remained deep in the crust of the earth. Had this magma reached the surface it would have become volcanic lavas.

The interior of the earth is hot and well insulated by the thick rock layers overhead. Magma that remains deep cools more slowly than surface lavas. This allows mineral grains to grow much larger than is seen in lava flows and accounts for why the rocks vary in texture, but not in composition. Together, igneous rocks can thus be divided between volcanic (cooled from magma that reached the

surface) or plutonic (after Pluto, the god of the underworld).

Gneiss is the most abundant of Connecticut rocks. Gneisses don't look much like sedimentary or volcanic rocks, but some resemble granite. They are often recognized by the presence of bands of light and dark minerals. Some bands appear to have flowed like magma, a clue they were formed under high temperature. Various gneisses are chemically similar to common igneous and sedimentary rocks, although their minerals may differ. Gneiss and related rocks are generally older than igneous and sedimentary rocks found in the same regions.

Connecticut gneisses were formed during the many orogenies that occurred in North America's past. These ancient mountain-building events pushed igneous and sedimentary rocks deep into the earth, to where tremendous heat and pressure transformed the rocks into gneiss, schist, and slate. Many outcrops in Connecticut reveal rocks in various stages of metamorphism, including forms intermediate between gneiss and granite or schist and mudstone. Metamorphism can be slow, but it is thorough, and must still be underway today, far below earth's greatest mountain ranges. All of the granites, pegmatites, and sedimentary rocks in Connecticut escaped this fate only because they formed after, or perhaps at the end of, the last orogeny.

Overview of the geologic history of Connecticut

The orogenies

Connecticut likely began to take shape as a small portion of Proto-North America, the remnants of which can be found in the state's northwest corner. These Proto-North American, metamorphic rocks formed during the Late Proterozoic, before and during the Grenville Orogeny of about one billion years ago. Whatever ancient mountains were created eroded by early in the Paleozoic Era.

In Early Cambrian time, the ancient continents were located far to the south. Between 500 and 460 million years ago, a shallow sea washed over erosion-leveled areas of Proto-North America and deposited thick layers of sand, clay, mud and fossil seashells. Minerals from the seashells contributed to the formation of limestone when the sediments later hardened. In places, this limestone was later metamorphosed into marble. This marble belt can be seen today extending from western and northwest Connecticut, through western Massachusetts and into Vermont.

Metamorphism and mountain-building events accompanied a series of collisions that continued to impact the eastern side of Proto-North America. One by one, in the crust beneath an ancient ocean called Iapetos, belts of arc-shaped volcanic islands and fragments of continents came to be sandwiched between Proto-North America and other large continental plates.

During the Taconic Orogeny of some 460 million years ago, Paleozoic terranes in westernmost Connecticut were pushed 40 miles inland. This belt of slates and metavolcanic rocks today forms the Taconic Mountains. The Green Mountains and Berkshire Mountains of Vermont and Massachusetts also extend through Connecticut east of Proto-North America and west of the Hartford Basin Terrane. Plate collisions of the Acadian Orogeny created the Iapetos Terranes and most of the mountainous rocks and structures of western New England more than 400 million years ago. The rocks of eastern New England arrived during the Alleghenian Orogeny some 260 million years ago and mark the Avalonian Terrane. Over the hundreds of millions of years since these mountain-building events, erosion has worked relentlessly to wear these mountains down as well, many down to their bedrock roots.

The super-continent of Pangaea

Hundreds of millions of years of continental collisions and mountain-building events worked to create all the world's continents later merged them into the super-continent known as Pangaea. Assembled in stages from 460 million to 270 million years before the present, the formation of Pangaea involved major shifts of ancient continents and volcanic island arcs across the globe. Near the center and far from the oceans lay the area that would become known as Connecticut.

As it was completed, Pangaea began slowly drifting northward. By the Late Triassic, the vast landmass had grown unstable. Heat built up in the mantle below, and convection currents began to stretch the crust apart along a zone where the Atlantic Ocean would eventually form to separate Africa from North America. Depressions in the crust, or rift basins, formed in regions around the new ocean on both continents.

The Hartford Basin

In the area of central Connecticut, ancient faults began to move again. The Connecticut rift basin, known to geologists as the Hartford Basin, began to sag along the eastern border fault separating it from the Bronson Hills Terrane, from north to south, all the way from our modern Long Island Sound to the borders of Vermont and New Hampshire. As this central area dropped down it formed a path that the present Connecticut River still follows over much of its course. At about the same time, the eastern and western regions of the state moved upward, encouraging stream and rain erosion that cut as much as five miles deep.

Sometime after the Early Jurassic, the strata in the basin began to tilt down along the eastern border fault. The resulting tilt, 15-20 degrees down to the east, is seen in road cuts throughout the valley today, and represents a feature of the

A rendering of Pangaea by paleontologist Paul E. Olsen
Courtesy Paul. E. Olsen

ancient world familiar to many.

The traprock ridges

East of the basin, the Bronson Hills grew more pronounced. Masses of boulders, stone and coarse gravels were swept from them and down into the basin. Beginning in the Late Triassic and continuing into the Early Jurassic, long periods of rapid sedimentation preceded and followed relatively brief periods of volcanic eruption. This pattern of deposition worked to fill the basin with its characteristic "layer cake" of sediments, including sandstones, mudstones and siltstones.

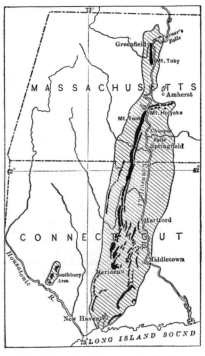

The Hartford Basin, including the Deerfield Basin of western Massachusetts
Courtesy, Connecticut Department of Environmental Protection

Remnants of ancient lava flows are seen today in the traprock ridges that extend nearly the length of the basin, from New Haven to Northfield, Massachusetts. To the east is a system of ridges formed by extrusive lava, or lava that once flowed across the surface of the valley, known as basalt. The eastern ridge system includes the Hanging Hills of Meriden and Totoket and Higby Mountains of the Metacomet Ridge. To the west is a smaller system that includes West Rock, Sleeping Giant and the Barndoor Hills. The western ridges are portions of intrusive magmatic sheets that hardened thousands of feet beneath the surface, between layers of sediment, during the first of the lava flows. They are made of dolerite, the intrusive equivalent of basalt. Because volcanic rocks are harder than surrounding sandstones, subsequent periods of erosion and eastward tilting have worked to leave the ridges standing out above the rest of the valley. Their raised western faces stand out dramatically as high cliffs of eroded, columnar basalt.

The Central Atlantic Magmatic Province

Many geologists once assumed that volcanic events like those known from the Hartford Basin occurred separately in many rift basins along eastern North America. More recent studies have shown that molten rocks that once flowed in Connecticut are exactly like Early Jurassic flows known from other rifts basins along eastern North America, such as the Culpeper Basin of Virginia, the Gettysburg Basin of Pennsylvania, the Newark Basin of New Jersey and the Fundy Basin between New Brunswick and Nova Scotia, Canada. All flowed across sediments of the same age and all have similar chemical compositions.

Geologists have determined that lavas erupted about the same time across a huge region of central Pangaea. This region, the "Central Atlantic Magmatic Province" (CAMP), has today come to be recognized as the largest volcanic feature on earth besides ocean crust. It stretches from the Canadian maritimes, along the eastern United States, south to Texas and Brazil, and from France to Liberia on the other side of the Atlantic.

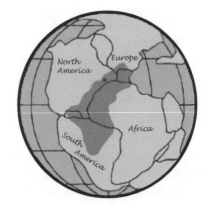

The Central Atlantic Magmatic Province (darkest area) of central Pangaea. Illustration by Greg McHone, after a map by Paul E. Olsen.

The Cretaceous

Toward the end of the Mesozoic Era, thick layers of

sediment collected on the newly formed continental shelf that extended along most of the eastern coast of North America, carried by ancient rivers that flowed from the Appalachian Mountains. Where sediment was deposited on land, the material formed the Coastal Plain, including the uplands of western Connecticut.

This widespread deposition of thick continental shelf and coastal plain sediments to the east was caused by uplift of the Appalachians during the Early Cretaceous, which according to some recent evidence resulted in a mile or more of vertical erosion. This uplift occurred all across the Northeast, and it is possible that mountain ranges such as the Adirondacks, Catskills, Berkshires and Green Mountains date from this Cretaceous uplift. The western and possibly eastern uplands of Connecticut appear to be involved, although today they are only hilly regions.

The Ice Age in Connecticut

Most people have heard of the "Ice Age" in which Connecticut (along with most of northern North America and Europe) was covered by great sheets of ice, or glaciers. The Ice Age in New England began about two million years ago (essentially the Pleistocene Epoch), but there were several—probably three and possibly more—times when the ice receded back into Canada, followed by new advances. Evidence concerning older ice advances is scanty, because most features and glacial sediments were destroyed by later advances. However, there are a few locations where two or three layers of glacial gravel each record different ice movement directions, as shown by different orientations of elongated cobbles and pebbles in the layers. Each of these layers might record different episodes or advances of ice.

The most recent advance came through about 24,000 years ago. That ice sheet spread rapidly reaching as far as the southern edge of Long Island and out into the present Gulf of Maine. Because it has been only 13,000 years or so since the ice left New England in the last recession, some geologists warn that the Ice

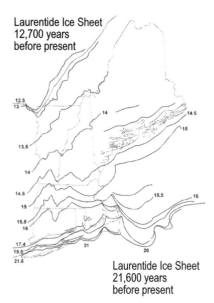

Laurentide Ice Sheet
12,700 years
before present

Laurentide Ice Sheet
21,600 years
before present

Laurentide Ice Sheet margins in New England. Numbered lines give positions of ice margins as glacial ice retreated (years before present). Courtesy of Byron Stone, United States Geological Survey.

Age might not be over, and the great glaciers could advance once again within a few thousand years. Such blocks of time are only the blink of a geologist's eye in the great calendar of earth events.

The end of the ice sheet is where the "conveyor belt" of moving ice melted and deposited all the detritus that it scraped up along the way. Whenever the ice sheet was stable for some period of time (perhaps a few decades to centuries), the materials dumped at the edges built up long ridges, called moraines. If these moraines contain some woody material mixed in, they can be carbon-dated. Work on end moraines in New England has resulted in maps that show where the edge of the continental glacier stood at different times in the past. Note that the farthest extent of the continental glacier left moraines that created our present-day Long Island, almost 22,000 years ago.

The thickness of this "Laurentide ice sheet" (named for the Laurentians of Quebec) must have exceeded the highest mountains at the peak of glaciation, perhaps over 6,000 feet. As the ice sheet melted later in the last cycle, scattered local glaciers were left along some mountainsides, with ice that did not overtop the mountains but only filled and flowed down the valleys in higher elevations.

Connecticut during the last Ice Age might have looked something like modern Greenland along the edges of its ice cap. At times, smaller valley glaciers in the eastern and western "highlands" flowed into larger glaciers. At other times, the entire land was hidden beneath one great ice sheet.

Where glaciers moved within valleys, the mountainsides were eroded into fairly steep slopes, while the valleys beneath the ice were cut deeper. Where the ice flowed over everything, the hills beneath the ice were rounded off, but the low areas were not significantly deepened. Valleys were dumping areas for glacial sediment called "stratified drift," in which sand and gravel have been separated into layers by lakes and streams melted from the ice.

The last glacier flowed right over most of Connecticut and then left, while local valley glaciers continued for a few thousand years in the higher mountain areas of northern New England. Sediment deposited from the melting glacial ice is called till (also called drift or hardpan) which mantles the bedrock of our hills. Often the bane of Connecticut gardeners, glacial till today forms a dense layer of all sizes of rocks as well as sand, silt, and clays just below our thin soils.

We can tell in which directions the last glaciers moved from grooves and scratches in bedrock surfaces. These grooves were made in bedrock by stones that were frozen in the base of the ice and scraped across bedrock as the ice moved. There are also places where there are more than one direction of grooves, indicating a change in the ice flow before it receded. Glacial grooves are very common all over New England, as is the smooth "polish" that the ice imparted to many bedrock surfaces.

The great moraines north of present-day Long Island were originally more continuous than today, so that when the continental ice sheet started to recede, fresh water filled what is now Long Island Sound behind a dam made of the long moraines. This former lake is known as "Lake Connecticut." At that time, so much water was still locked up in continental glaciers that sea level was much lower. Lake Connecticut formed while the Atlantic Ocean was many miles out from today's shoreline. As the ice melted and sea level gradually rose, the ocean moved closer and eventually the freshwater in Lake Connecticut came to be replaced by seawater.

Glacial Lake Hitchcock
Courtesy of Dr. Julie Brigham-Grette, Dept. of
Geosciences, University of Massachusetts, Amherst

As the ice melted farther to the north, meltwater filled lowland areas and numbers of glacial lakes formed. Among these was a great lake that has since become known as Glacial Lake Hitchcock, named for its discoverer, Edward Hitchcock. Lake Hitchcock was very long, stretching all the way from a natural dam formed at Rocky Hill, Connecticut, north to Lyme, New Hampshire, and probably up several side valleys as well. The remnants of the Rocky Hill dam, a delta of sand and gravel formed in a previous glacial lake, can still be seen. It has been quarried for many years, but portions of its level top remain. Thin layers of mud deposited in Lake Hitchcock, known as varves, also survive. Geologists have learned to count varves almost like tree rings and they have been used to precisely determine the ages of events that affected the lake. Varves are also especially useful for the records they provide of climate.

Other glacially dammed lakes, such as Lake Albany and many smaller lakes, formed in upland regions. An arm of the ocean reached up the present-day St. Lawrence River of Quebec and into what is now the Lake Champlain Valley, to form the "Champlain

Sea." Fossil remains of oysters, whales and other marine animals have since been found in that valley.

Immediately after the ice left, plants and animals quickly colonized newly thawed areas, even as it remained very cold. Amazingly enough, nearly complete skeletons of large animals that roamed the tundra-like landscape are still discovered preserved in lake muds from 10,000 years or more in the past.

Earthquakes in Connecticut

There are plenty of faults in Connecticut (although most are obviously very old), and we know that earthquakes other than those caused by landslides or explosions always occur by the movement of a fault (although it might remain hidden from the surface). And yes, we do sometimes have small earthquakes in and near Connecticut. So, should we worry?

The fact is that no earthquake in our state has ever been related to a specific fault marked on any map. Faults that have moved probably don't move enough to become obvious. Connecticut earthquakes are more likely to be topics of conversation than anything else.

An interesting account of the largest earthquake known in Connecticut is found in *Seismicity of the United States,* 1568-1989 (Revised), published in 1993 by Carl W. Stover and Jerry L. Coffman as U.S. Geological Survey Professional Paper 1527, United States Government Printing Office, Washington D.C. The earthquake occurred near Moodus, in the Town of East Haddam on May 16, 1791 around 1 p.m. The earthquake's size is estimated from historical descriptions to be Intensity VII on the Mercalli scale, which might be around magnitude 5 on the Richter scale. According to accounts by Native Americans, this region of East Haddam, near where the Salmon River meets the Connecticut River southeast of Middletown, has been the scene of local disturbances long before the first European colonists arrived. The name Moodus is taken from the name Native people gave the place.

They knew it as Machimoodus, or "place of noises." The noises were described as similar to distant thunder or cannon fire, and were somehow caused by earthquakes generally too small to feel.

The earthquake of May 1791 began with two heavy shocks in quick succession. Stonewalls were shaken down, chimney tops broken off and latched doors were thrown open. A fissure several meters long formed in the ground. Thirty lighter shocks followed. Through the night more than one hundred shocks were counted. The largest were felt as far away as Boston and New York City.

Today, tremors are monitored by the Weston Observatory of Boston College, Massachusetts. The Observatory monitors and maintains devices known as seismographs to measure earthquakes throughout

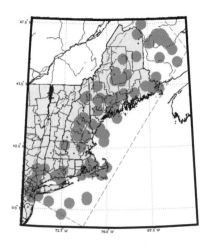

A map showing areas of New England where there is a chance of an earthquake occurring (gray areas) based on an analysis of the earthquake activity of magnitude 2.7 or greater from 1975 to 1988. According to a Weston Observatory analysis of earthquakes that occurred between 1989 and 1998, there is a 66% chance that the next earthquake of magnitude 2.7 or greater in New England will occur in one of the gray-shaded areas.

Map: Dr. Alan L. Kafka, Weston Observatory of Boston College.

New England. According to Dr. Alan L. Kafka, there is a 66 percent chance that the next earthquake of magnitude 2.7 or greater will occur in one of a number of areas of New England that the Observatory monitors.

That would be fairly small as earthquakes go—and the risk of damage from earthquakes anywhere in Connecticut is considered by most experts to be minimal. Most of the potential earthquake zones in Connecticut are clustered between Moodus and New Haven.

Another zone is found in westernmost Connecticut, around the greater New York City area. Outside the state, earthquakes may occur in southeastern Massachusetts and southern New Hampshire. The state of Maine is surprisingly active as well.

A Geological Column for Connecticut (by Greg McHone)

Years in Millions	Eon	Era	Period	Epoch	Geological & Evolutionary Events
0.011	Phanerozoic	Cenozoic	Quartenary	Holocene	Humans arrive
1.8				Pleistocene	Dramatic changes in climate; **Ice sheets** cover & uncover CT
5			Tertiary	Pliocene	Marine coastal sedimentation
23				Miocene	Erosional interval
38				Oligocene	Erosional interval
54				Eocene	Semi-tropical plants in Vermont
66				Paleocene	Erosional interval; **Mammals expand**
146		Mesozoic	Cretaceous		Shallow sea covers coastal New England; Uplift & erosion inland; **Birds expand**
201			Jurassic		Atlantic Ocean opens, south flowing rivers (Ct., Thames, etc.) develop; Climate grows wetter; CAMP Lava flows; **Dinosaurs expand**
250			Triassic		Rifting of Pangaea begins; sedimentation of Hartford Basin; SW rivers flow in semi-arid climate; **Primitive dinosaurs appear**
286		Paleozoic	Permian		**Pangaea completed; Alleghenian Orogeny:** Alleghenian mountain building; Stony Creek granites, eastern pegmatites
325			Pennsylvanian		Coals deposited in coastal swamps; Alleghenian Orogeny begins
360			Mississipian		Passive margin sedimentation
410			Devonian		**Acadian Orogeny:** Sedimentation in eastern CT, granites, western pegmatites melted by burial, others metamorphosed; **Early amphibians**
440			Silurian		Taconic highlands eroded; Land plants
505			Ordovician		**Taconic Orogeny:** marine muds & sands buried and metamorphosed
544			Cambrian		**Cambrian Explosion**; Fossils seashells; Carbonate depositions on passive margin
650	Precambrian		Neoproterozoic		600-550: Late phase Iapetan rifting
					750-700: Early phases Iapetan rifting
900					
			Mesoproterozoic		**Grenvillian Orogeny** (1,100-950)
					Oldest rocks in Connecticut
1,600					
			Paleoproterozoic		Atmospheric oxygen increases
2,500					

4.6 billion years (Precambrian)

The earth, along with the other planets and the sun, coalesces from a large cloud of gas and dust that form the solar system. Our present moon splits off from the early earth within a few hundred million years, perhaps from a planetary collision.

4 billion years (Precambrian)

Less dense rocks of the earth's crust separate from the heavier mantle and iron core. The surface cools enough for the atmosphere to stabilize and for water to condense from steam from volcanoes. Unlike most planets, earth's volcanoes continue to produce gases that enrich our atmosphere and allow plants to flourish.

3.8 billion years (Precambrian)

Life appears, indicated by a "light carbon" isotope in the oldest sedimentary rocks known (from Greenland). Fossils were not well preserved, but tiny spherical marks may be from bacteria. The air is rich in nitrogen (N) and carbon dioxide (CO_2) with some methane, water vapor and other gases, but very little oxygen (O_2) is present.

3.5 billion years (Precambrian)

Layers of simple bacteria and algae-like cells work together to form the stromatolites (dome-shaped structures) in shallow warm ocean muds. Fossils formed from stromatolites in Australia survive as the oldest known.

2 billion years (Precambrian)

There are signs, such as oxidized iron minerals that collect in sedimentary layers, that O_2 is becoming more abundant in the atmosphere. Eukaryotes (cells with a membrane-bound nucleus, the ancestors of all animals) appear.

1.4 billion years (Precambrian)

The oldest rocks in Connecticut, in the southwestern area of the state, recrystallize from even older rocks during a collision of tectonic plates called the Grenville Orogeny. These are similar to metamorphic rocks in Vermont and New York that today lie beneath sedimentary layers of Proto-North America.

1 billion years (Precambrian)
Photosynthesizing organisms (algae and plankton) thrive in shallow seas. These plants metabolize CO_2 and release O_2 into the air as a by-product. The Grenville Orogeny is ending in eastern Proto-North America, which includes westernmost Connecticut. The super-continent of Rodinia forms and dominates the map of the world for the next 400 million years.

600 million years ago (Precambrian)
Marine photosynthesizing organisms remove so much CO_2 from the air that the climate becomes very cold, causing a massive "snowball earth" ice age. Tropical oceans are frozen beneath a mile of sea ice. Volcanic activity eventually renews CO_2 and the earth swings wildly back, to a hot, ice-free climate. Rodinia breaks into several large continents. Simple marine animals, without hard parts, become larger and very diverse. Rocks of eastern Connecticut, known as the Avalon Terrane, form in a volcanic zone along the western margin of Gondwanaland, one of the large continents that are moving away from Proto-North America.

545 million years (Cambrian)
The "Cambrian Explosion" of marine animals is recorded by fossils of a huge variety of animals, including corals and sponges, worms, clams, arthropods, etc. Many animals inhabit the shallow, calcareous-mud bottom of warm shallow oceans along the edges of the continents, including westernmost Connecticut. Early fish appear, the ancestors of all modern vertebrates (animals with backbones) late in Cambrian time.

480 million years (Ordovician)
A mass extinction of unknown cause marks the end of the Ordovician. Western Connecticut rises as ocean volcanic rocks are

pushed into the ancient continent during the Taconic Orogeny. The eastern two-thirds of the state remain under an ancient ocean called Iapetos.

425 million years (Silurian)

Simple wetland and shoreline plants colonize the land. Scorpions and other arthropods also move onto the land, aided by oxygen levels in the air now above ten percent.

395 million years (Devonian)

The first vascular plants (plants with specialized structures that conduct fluids and add structural strength) appear. Ferns, reeds and horsetails, rapidly evolve. Land-dwelling insects grow common. The first amphibians evolve from lobe-finned fish. They are able to breathe air and crawl on land, but return to the water to lay eggs. The Iapetos Ocean is destroyed as its crust is pushed into the continent (the Acadian Orogeny), causing the seafloor muds to metamorphose into gneiss and schist. Granites rising in the crust of west-central Connecticut melt. Volcanic island arcs and distinct regions of the Iapetos Ocean form the geologic zones of New England known as terranes. At the end of the Devonian, an extinction eliminates about half of all the world's living species.

330 million years (Carboniferous)

Tall coniferous trees appear. Large flying insects (dragonflies, roaches, cicadas) appear. Frogs develop. Early reptiles with dry, scaly skins appear, which can lay eggs on land. Great swamps form coal deposits around the world, including along the margin of the continent of Gondwana. Gondwana moves westward to collide with our new continent during the Alleghenian Orogeny, forming the Avalon Terrane of southeastern New England. The collision completes the Appalachian Mountains, which extend over much of Connecticut and the central part of the new super-continent of Pangaea.

270 million years (Permian)

Very high O_2 levels (up to 40 percent, compared with 21 percent today) allow huge insects to fly despite primitive respiratory systems. Reptiles (including the ancestors of modern mammals) become diverse and large. CO_2 has dropped again, and an ice age occurs across polar regions. The greatest mass extinction in history wipes out 90 percent of living species at the end of the Permian 250 million years ago. The extinction may have been caused by vast lava flows and volcanic gases or a bolide impact. The Appalachian Mountains still reach high, but are rapidly eroding.

225 million years (Triassic)

Life takes tens of millions of years to recover from the Permian-Triassic extinction. Reptiles grow large and diverse to crawl over land, swim the oceans and glide through the skies. Large forests of conifers, cycads, gingkos, and tree ferns mark a warm, but dry global climate. Mass extinction marks the end of the Triassic, again possibly due to meteor impacts or a series of volcanic events which flooded central Connecticut. This eliminates the large land reptiles and clears the way for dinosaurs to expand.

150 million years (Jurassic)

Huge plant-eating dinosaurs called sauropods walk on land. Large marine reptiles dominate the oceans. The first true birds evolve from theropod dinosaurs. Animals and plants become more diverse. The Atlantic Ocean has opened with a shoreline east of Connecticut, while once tall peaks erode rapidly.

120 million years (Cretaceous)

Flowering plants evolve and become as abundant as trees, grasses, herbs, shrubs, etc. Bees and butterflies follow. More birds and mammals evolve. Sea levels rise until southern Connecticut is beneath the shallow western Atlantic Ocean. The last of the "big five" mass extinctions marks an end to the age of dinosaurs, as well as all flying reptiles and the remaining marine reptiles, and the Cretaceous. This time, both a large volcanic lava event in India and a huge meteor impact may be responsible. The largest crater known on earth, buried in the western Yucatan Peninsula and Gulf of Mexico, provides compelling evidence of an asteroid impact.

50 million years (Tertiary)

Mammals and birds inherit the earth and, without competition from dinosaurs and pterosaurs, rapidly expand. Early primates evolve, and other mammals return to the sea as whales. The climate cools, although it still feels subtropical in Connecticut. The land slowly rises and the Cretaceous beach sands are washed away from southern Connecticut.

1 million years (Quaternary)

The Ice Age covers northern continental areas as well as Antarctica with sheets of continental ice up to a mile thick. Early humans flourish, learn to use tools and fire, and gradually come to dominate all other living things. The movements of thick ice sheets over Connecticut erode our mountains and valleys and later fill many of the valleys with thick sand and gravel deposits.

12,000 years (Recent)

The great sheets of ice melt back into Canada and most of the large Ice Age mammals of Connecticut die out. The first Native Americans move into Connecticut. The climate rapidly warms to its present state, and plants and animals familiar to us today inhabit our landscape.

The Metamorphic Terranes 4

Photo: Greg McHone

Most of Connecticut is made of rocks with long and complex histories, products of plate tectonics, known as metamorphic rocks. After powerful evidence of the movement of large sections of the earth's crust was published in the 1950s, geologists working in eastern North America constructed new explanations for these complex rocks and structures all along the Appalachian Mountain belt. Their ideas involved radical proposals for collisions of ancient oceans and continents, each one causing faulting, folding, and metamorphism in different sections of New England, ideas that have since come to be known as "New Global Tectonics."

Adapting to the new concept of global tectonics proved difficult for many geologists. For some, it meant reevaluating a career's worth of work. It fell to a few, farseeing geologists to construct a totally new explanation of the geological history of the Appalachians, including John Dewey of the State University of New York at Albany, Hank Williams of Memorial University in Newfoundland, and John Rodgers of Yale University. The model they proposed gained acceptance because it was able to explain features of the local region that had puzzled geologists since the 1800s. They realized that elongated regions within the wide mountain belts, which included nearly all of New England as well as Atlantic Canada and the southeastern states, were of different ages and feature different rocks and rock structures that testify to their unique geological histories. Among the more remarkable aspects of this new model was the notion that many of the metamorphic rocks in these mountain belt zones were formed from sand and mud that had been deposited at the bottom of ancient oceans—and were far older than features associated with the modern Atlantic Ocean.

Looking at things from this new perspective, these geologists realized the histories of the rocks of many of these mountain zones began long before the Atlantic existed. They had, in fact, originated as parts of other continents and oceans and were once located in areas of the globe far from where they ended up. Movements of the earth's plates during Paleozoic time (and possibly earlier) led to continental collisions (or continental-continental convergences) that caused a series of mountain building events or orogenies. During each orogeny, rocks along collision zones were compressed together until they folded in great folds or cracked to form faults. At times, slices of rock slid over or under each other as much as 50 miles or more.

As earth was compressed, the crust thickened. Great masses of rocks that were once at or near the surface were buried ten miles or more deep. At those depths, the pressure and heat built up to tremendous levels—until the process of metamorphism worked to transform the rock. In metamorphism, rocks are transformed by the growth of new minerals and fabrics. Some may even have melted partially to form large bodies of granite or pegmatite. Few were left looking anything like the sediments or rock materials they had been before being subjected to metamorphism.

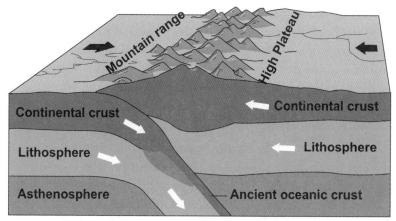

A schematic view of "continental-continental convergence"
from This Dynamic Earth, United States Geological Survey

During and after each collision, mountains were raised high. The higher areas of crust were elevated, the more quickly the surface eroded, as much as two to three feet every thousand years. After some millions of years, the mountains were eroded so much that metamorphic rocks that had been formed deep beneath them came to be exposed. Several miles of the upper layers must have been eroded away to expose these deep orogenic rocks—and these are the rocks we find in abundance across most of Connecticut today.

The crustal pieces of these ancient plates define the terranes of Connecticut. The term is a new use of an old word and has a different meaning from the word "terrain," which describes a distinct region or landform feature of the earth's surface. Connecticut is comprised of about nine terranes, depending on how they may be defined. Boundaries between terranes are often (but not always) major faults along which crustal pieces slid into one another. Older boundaries may themselves become folded and faulted from subsequent orogenic collisions.

Nearly all are made of metamorphic rocks, of which gneiss and schist predominate. The central lowlands of the state are a notable exception. Geologists know the central rift basin as Newark Terrane, a name that recognizes its similarity to basins found up and down eastern North America and the sedimentary strata that separate these relatively younger features from older, metamorphic rocks surrounding them.

The Metamorphic Terranes

The northwestern corner of Connecticut preserves a small portion of Proto-North America, an ancient continent that previous to 460 million years ago was smaller than North America is today. Beginning some 1.4 billion years ago, a series of orogenies occurred around this portion of Proto-North America, then part of a super-continent called Rodinia, but these events remain somewhat mysterious. Some geologists suspect that plate tectonic processes may have functioned differently during the early history of the earth. The interior of the planet was hotter then and continents may have moved faster than they do today.

Connecticut began to take shape before and during the Grenville Orogeny of about a billion years ago. The metamorphic rocks of the portion of Proto-North America that remains in Connecticut were formed during these Late Proterozoic events. About 250 million years later, Rodinia broke apart and Proto-North America became one of several continents in motion across the earth.

Whatever ancient mountains were created during the Grenville Orogeny had eroded nearly flat by early in the Paleozoic Era. Between 500 and 460 million years ago, a shallow warm sea spread over the lowered, level areas of Proto-North America. Thick layers of sand, clay, and mud were deposited, including many fossils of marine, shelled animals. Minerals such as calcium carbonate in the shells contributed to the formation of limestone as the mud later hardened. The limestone, in turn, was later metamorphosed

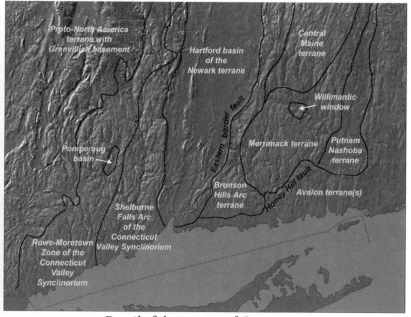

Detail of the terranes of Connecticut
Illustration by Greg McHone

into marble. Today, this "marble belt" extends over a portion of
New England, from the northwest Connecticut hills and through
western Massachusetts and Vermont. Local marble has been
quarried in these states for many years.

To the east of Proto-North America was an ocean called
Iapetos (named after the father of Atlas of Greek mythology). In
and around the Iapetos Ocean, arc-shaped belts of volcanic islands
and small pieces of continents were sandwiched in between the
larger continents. Plate tectonic motions moved the plate that held
Proto-North America toward an even larger continental plate called
Gondwana, and one by one, the smaller ocean crust, volcanic arcs,
and continental plates in between were pushed into the eastern side
of Proto-North America. These collisions caused orogenies, and
each one contained rocks metamorphosed deep within a mountain

belt. Today, these masses of recrystallized rocks are exposed in Connecticut as terranes, with boundaries that mark huge faults where the plates slid over, under, or along one another.

During the Taconic Orogeny of about 460 million years ago, another Paleozoic terrane was pushed 50 miles or more over eastern Proto-North America. The collision raised mountains thousands of feet tall. Only a tiny piece of this belt of slates and meta-volcanic rocks that forms the Taconic Mountains remains in the northwestern corner of Connecticut.

The Connecticut Valley Synclinorium Terrane, including the eastern areas of the Green Mountains and Berkshire Mountains of Vermont and Massachusetts, extends farther to the east of Proto-North America in Connecticut, to the west of the Hartford Basin Terrane. In fact, this region of metamorphic rock forms the "basement" terrane beneath the Mesozoic sedimentary and volcanic layers of the state's central lowlands. Metamorphic minerals in this western terrane are about 380 to 400 million years old, and were formed during the Acadian Orogeny. Many granites and pegmatites in western New England were formed near the end of this mountain-building event.

A large volcanic island arc with some sedimentary materials also collided with Proto-North America during the Acadian Orogeny. Later collisions re-metamorphosed these rocks, however, and pushed them westward. As a result, they show younger mineral ages, closer to 300 million years old. This section of Connecticut now remains as a belt along the eastern side of the Connecticut Valley known as the Bronson Hills Terrane. Other plate sections, once within the Iapetos Ocean, lie farther east and are named the Merrimack, Central Maine, and Putnam-Nashoba Terranes.

All of these eastern Iapetos Terranes were greatly affected by the last great plate collision around 300 million years ago. The resulting Alleghenian Orogeny takes its name from a mountain belt in Pennsylvania. During this event, a long and independent plate, called Avalonia, collided with the masses of Iapetos Terranes that had collected along eastern North America. The Avalonian Terrane itself is a mix of much older (600 to 700 million years) formations of continental and volcanic rocks.

The force of this collision pushed much of eastern New England westward. Avalonia slid at a shallow angle beneath the eastern Iapetan Terranes at least as far as the Bronson Hills. This event also caused granites and pegmatites to be melted out of the older rocks. Some of their newly formed minerals include pink-colored feldspars, and handsome pink Avalonian granite is still quarried today in Stony Creek, an area of Branford.

Making metamorphic rocks

It has been a puzzle for thousands of years: How was it possible that rocks that looked curiously like seashells were sometimes found on mountainsides, high above sea level? The great Renaissance artist and philosopher, Leonardo da Vinci wrote about stones he thought were seashells in the mountains of Italy in the 1400s. Nicholas Steno, of course, was inspired to pioneer basic principles of geology looking at strange stones he found in Italian mountains more than a century later.

Some believed that it was possible marine animals might somehow have been transported to high ground in past floods, possibly even the Great Flood described in the Bible. Leonardo, however, came to understand that "where there is now land, there once was ocean." Among his many great contributions to science, Leonardo centuries ago conceived of how the earth had actually been made over many times during its long history.

We know now that certain geological formations, such as the marble ridges of Connecticut, started out simply as mud beneath shallow seas just as Leonardo had suggested. Through a biological process known as biomineralization, certain marine animals are able to extract minerals from seawater and supply them to their cells to grow hard shells. Among the minerals commonly found in seashells is calcium carbonate or lime. As shells collected in the mud at the bottom of ancient shallow seas the mud was enriched with lime, just as shallow areas of warm seas in the Caribbean are today. These lime-rich muds later hardened to form limestone.

Limestone can be changed into marble if it is subjected to some heat, above several hundred degrees Fahrenheit, and pressures of thousands of pounds per square inch. Under these conditions, calcium carbonate minerals (mainly calcite) will slowly grow to form new, larger crystals in the rock in a process called recrystallization. Recrystallization can change (meta) the entire form (morph) of the rock. Look closely at pieces of limestone and marble and you can see differences in the sizes of the grains as well as new patterns of grains caused by deformation of the rock material.

Conditions of high heat and pressure are known to occur naturally miles deep in the crust, but until recently geologists had no way to account for how the seafloor might be buried deep enough to be exposed to the kinds of heats and pressures required to change limestone into marble.

Before the concept of plate tectonics was developed, geologists thought that sediments deposited in long-lived river deltas along continental margins might slowly reach great depths, places such as where the Mississippi River deposits millions of tons of mud into the Gulf of Mexico. It was thought that given time enough sediment could accumulate to cause depressions in the ocean floor. The lower layers of these sediments would eventually reach depths where they would slowly be metamorphosed. Such areas are now known as "geosynclines." They can form beneath sedimentary basins. The problem with this model is that it does not explain how continental dry-land sediments and igneous rocks can become metamorphosed or why we commonly see metamorphic products in mountain belts.

New Global Tectonics provides better answers to the mysteries of seashells on mountaintops and metamorphic rocks. It turns out that what can be built up can also be forced down. Crust in collision areas may be pushed up into high mountain ranges or driven down to great depths where it is metamorphosed.

Where plate tectonics cause continental crust and/or ocean crusts to collide, the side with heavier rocks (usually ocean crust) tends to be forced under the lighter, less dense plate. As this happens, the crust in the collision zone becomes much thicker. It may become folded and faulted. Thick slices of crust may slide into one another like huge wedges. In all, crust along collision zones may double in thickness, increasing from perhaps 20 miles thick to as much as 40 miles thick, while tall mountains rise above the surface.

Collisions can also work to force sediments and rocks once at shallow levels (especially ocean silt and mud) down to the great depths, where they recrystallize. These new metamorphic rocks may become folded and faulted, perhaps even partially

Photo: Greg McHone

melted. Erosion can work to wear away at the steep surface relief of young mountain ranges, eventually causing even deeply buried metamorphic rocks to be exposed. Many of the metamorphic rocks in Connecticut were formed deep beneath former mountain ranges and exposed after great periods of erosion.

A similar process is underway today deep within the Himalayan Mountains of Asia, just as happened long ago during the formation of the Appalachian Mountains of eastern North America. Mountain-building events, or orogenies, that are part of Connecticut's geologic history explain our abundance of metamorphic rocks, and our scattered granites and pegmatites as well. Rare soapstone and amphibolite rocks are pieces of volcanic sections of the ancient Iapetos Ocean crust, and its thick ocean muds were completely transformed into our abundant schists and gneisses.

Studying and interpreting these features challenge even the most talented field geologists. A metamorphic petrologist must be able to decipher many different minerals that formed under various temperatures and pressures, as well as measure and interpret complex structures and fabrics created by the tectonic events of plate collisions. Luckily, powerful tools are available at university laboratories, where chemical compositions and radiometric ages of microscopic minerals can be determined. Field geologists and their students regularly meet to communicate their findings (and defend or dispute their interpretations). After several decades of hard work, our knowledge and understanding of earth history has been greatly improved.

Photo: Greg McHone

Avalonian Gneisses in Salem

Some day Route 11 will be finished in southeastern Connecticut, but before then you may have an opportunity to walk along some road cuts where the highway will eventually run.

How to get there: From Route 2 in Colchester, take Route 11 south to its end and turn left (east) onto Route 82, toward Salem Four Corners. Go under the unused Route 11 overpass and park in a gravel lot east of the state highway maintenance buildings, on the right (south) side of Route 82. Take the trail from the gravel parking area south, and follow the future pathway of the north-bound lanes of Route 11. It is about ¼ mile to road cuts along this highway grade. As always, watch your footing and do not climb the cliff face.

What you will see: This area is along the contact between gray-

Honey Hill Fault slope
Photo: Greg McHone

layered Rope Ferry Gneiss and pink granitic Hope Valley Gneiss, both of which are easy to see in the road cuts. These formations originated at least 600 million years ago along the margin of Gondwana in an area that is now part of western Africa, but which later broke off to form the narrow but long Avalonian continent. Its collision with our region about 300 million years ago created newer minerals and structures that overprint the older features, which for a time caused some confusion about the history of these rocks.

The coarse pink granitic gneiss was probably an igneous granitic rock called alaskite before it was metamorphosed. Alaskite is a plutonic rock that contains alkali feldspar, quartz, and muscovite, but few dark minerals. The dark gray layers of gneiss that you see in the rock face may represent volcanic ash or lava flows that have also been thoroughly metamorphosed. Although their minerals were nearly all re-made (recrystallized) because of metamorphism, the chemical compositions of these rocks can reveal the original "protolith" identities.

As you return to the parking area, you have a good view across the valley of a long low ridge called Honey Hill. The ridge is made of much younger formations of the Merrimack Terrane, and the great fault that marks its contact with the Avalonian rocks behind and beneath you is called the Honey Hill fault. Like most faults it is not well exposed, but highly deformed banded structures caused by the fault motion along the base of Honey Hill are recognized on Route 11. The Avalonian Terrane slid along this fault far to the west beneath the Iapetos Terranes, and the heat produced by this deformation affected rocks all across eastern Connecticut.

Things to discuss: Pick up samples of the gray and pink gneisses. Can you see the differences that allow these two rock formations to be easily distinguished and mapped? Why do some people say that Avalonia is "a piece of Africa" left behind when the Atlantic Ocean opened up?

Photo: Greg McHone

Brimfield Schist, Amphibolite, & Aplite
Devil's Hopyard, East Haddam

A major formation of the Merrimack Terrane is the Brimfield Schist, named after a town in Massachusetts where it is also exposed. The Brimfield Schist is notorious for adding sulfur and iron to private wells due to a high pyrite content, but it also contains some interesting volcanic materials that were metamorphosed into amphibolite, as well as granitic dikes. All of these rock types are displayed below the scenic waterfall at Devil's Hopyard State Park in East Haddam.

How to get there: From the southern end of Route 11, head west a few miles on Route 82 to the intersection with Hopyard Road, turn right, and travel north to the parking lot along the Eight Mile River below the falls. From the west, you can reach the park by crossing the "swinging bridge" into Goodspeed Landing (East Haddam

Village) and driving eastward on Route 82. Turn left (east) onto Mt. Parnassus Road and travel several miles, pass Millington Green, and turn right onto Hopyard Road to get to Devil's Hopyard.

This handsome old state park was constructed by the Civilian Conservation Corps during the 1930s. Its quirky name is derived from a farmer named Dibble, who raised hops nearby for making beer (his name was distorted for local humor). To improve the story, the Devil has supposedly left a hoof mark by the falls. Take the trail along the west side of Eight Mile River until you are below the waterfall, and branch off to the outcrops below the falls. Take care to not slip on wet rocks!

What you will see: The Brimfield Schist is usually rich in biotite mica, but where the mica is replaced by dark hornblende (a common mineral of the amphibole group), the rock is called amphibolite. In the outcrops near the water, you can see a more-massive, solid rock with blocky shapes of hornblende crystals. It is thought that amphibolite can be derived from volcanic basalt through metamorphism. A dike of granitic rock called aplite (essentially a fine-grained version of pegmatite) cuts across the rock. Because it is straight, the dike must be much younger than the amphibolite schist, which has folds and foliations that were probably created in the Acadian orogeny. In contrast, the dike flowed into a fracture at the end of the Alleghenian orogeny, or possibly at a later time in which there was igneous activity.

Things to discuss: Can you guess some possible ages (relative or absolute) for these rocks, based on their structures and crosscutting relationships?

Photo: Greg McHone

Terrane Boundary in Deep River

Major terrane boundary faults are almost never exposed. However, it appears that one such spot exists in southern Connecticut, which was discovered by Dr. Robert Wintsch in the 1980s. Luckily it is in a public park, right next to a parking lot!

How to get there: From Route 9 take Exit 5, follow Route 80 eastward to Route 154 in downtown Deep River, and turn right (south). After about 0.4 miles, turn left onto Southworth St. and then immediately left into a parking lot between the highway and a playing field at the town park. Park and walk onto the small ridge toward the highway.

What you will see: The fire department of the Town of Deep River kindly washed out a small trench across the terrane contact years ago, although this has filled in somewhat since. Late Proterozoic Rope Ferry Gneiss (Avalon Terrane) is present on the eastern side, and dark granular schist of the Ordovician Tatnic Hill formation (Merrimack Terrane) is on the western side. The contact zone is intruded by several small bodies of white to pink pegmatites, which in places are broken up and "smeared" into lenses along the tectonic foliation. This is a "ductile" fault, which moved under such high heat and pressure that the rocks flowed and recrystallized into thin layers or bands. A "brittle" fault under lower pressure would appear as a sharp break with broken-up rock material along it, not what we have here.

In addition, several glacially-smoothed pavements (with glacial grooves) are exposed because the soil is quite thin.

Things to discuss: An outcrop like this might cause some controversy, as geologists want persuasive evidence about the interpretation of such an important location. Can you think of some reasons why it might not be obvious that this place is an exposure of a major terrane boundary? What evidence might be collected as conclusive proof?

Photo: Greg McHone

Collins Hill Schist at Cobalt Mines, East Hampton

Several small ore bodies of cobalt, lead, silver, and gold occur in the dark meta-volcanic schists of the Bronson Hills Terrane. These are vein deposits near Mesozoic faults, which apparently provided a pathway for hot mineral-rich fluids to rise. No mines operate today, but this area is reputed to be where Governor Winthrop secretly mined gold in the 1660s.

How to get there: From Middletown, follow Route 66 east through Portland to the village of Cobalt, where it intersects with Route 151. Or you can follow Route 66 westward from Exit 13 off Route 2 through Marlborough and East Hampton to Cobalt. Turn north at the light onto Middle Haddam Road, continue through the intersection and up the hill, and turn right onto Gadpouch Road into Meshomasic State Forest (one of our oldest state forests).

Park in the small lot on the right side near the driveway to the ranger's headquarters. Walk to the fenced in area in the woods above Mine Brook. Caution: Do not attempt to climb the ruins or enter any mine tunnels.

What you will see: Cobalt was mined here during the Revolutionary War and again in the mid-1800s, although it is not likely that the mine was very profitable. The fence surrounds an old shaft, and a substantial stone mill foundation is just to the south. The mine owner in the 1770s also built a distillery nearby, both to encourage his miners and to aid thirsty soldiers.

The gold and cobalt-nickel minerals at Cobalt occur in mineralized zones that appear to be related to an east-west fault zone. The gold occurs along with pyrite and native bismuth in small fractures within quartz-arsenopyrite veins that are several feet wide. In Mine Brook east of the old shaft you can see some of these white veins of quartz, within which are silvery-white metallic crystals of arsenopyrite (arsenic-iron sulfide). Cobaltite is a closely related mineral.

The Collins Hill Schist is rich in hornblende, which as in amphibolites may indicate a volcanic basalt origin. The metal-sulfide minerals may also be volcanic in origin, although they later moved into veins in the rock during Mesozoic tectonic events. The schist also contains biotite mica and small purple garnets, which tend to weather out of the rock like tiny beads.

Things to discuss: Given that the cobalt is mixed with arsenic and sulfur in these minerals, what environmental problems might result from indiscriminate dumping of these mine wastes?

Photo: Greg McHone

Canterbury Gneiss at Wolf Den, Pomfret

As you might guess from their massive textures and hard minerals, gneisses often have parental materials of plutonic rocks such as granite. However, they may also be formed from volcanic equivalents of granite such as rhyolite ash flows. Similar parental chemical compositions result in similar minerals to be formed by metamorphism. The Canterbury Gneiss is widespread in eastern Connecticut, and at this park it is associated with an historical site.

How to get there: Wolf Den is in the Mashamoquet State Forest in Pomfret. You can get there by heading west on Route 101 from Exit 93 off I-395 in Killingly. Or, head east on Route 6 from Willimantic to Brooklyn, turn left (north) onto Route 169, and left (west) onto Route 101. Soon after Route 101 merges with Route

44, turn left (south) onto Murray Road, and after a mile or so make a sharp left onto State Park Road. In less than ½ mile there is a right turn onto a side road, which dead ends at a picnic area and trailhead. Walk down to trail to the Wolf Den.

What you will see: The numerous rounded boulders are of Devonian Canterbury Gneiss, a light gray, medium-grained, lineated granitic gneiss. Lineations are thin lines formed by minerals in the rock, which represent a special deformation fabric. Some metamorphic minerals grew in response to the slow movements of the rock material, and so geologists always take care to measure lineations wherever they are found. The gneiss has the right composition for a granitic origin but mineral textures better related to volcanic materials such as ash.

Glaciers at the end of the Ice Age rounded these large boulders of gneiss, and moved some of them as well. Natural fractures called joints have opened up from frost and weathering, creating pockets and the small cave called Wolf Den. This is where Israel Putnam, a hero of the wars of the eighteenth century, killed what may have been the last wolves (mother and pups) in Connecticut. He had to crawl head first into the cave to shoot them, a brave if perhaps regrettable action.

Things to discuss: Why is the soil so thin on this hillside? Where do roots go that hold up trees in this forest? Could this gneiss be a good building material?

Photo: Nancy McHone

Ratlum Mountain Schist at Brett Woods, Fairfield

The ancient Iapetos seafloor had large sections of mud rich in clay. The chemical elements in clay are good for forming mica upon metamorphism, and if iron and magnesium are also present it will be the dark mica called biotite. Otherwise muscovite (a white mica) can form, or perhaps both minerals will be present. The mica grains are flat and tend to be aligned in thin layers, which creates the texture of schist. Examples of this rock type are found in many places, including southwestern Connecticut.

How to get there: From the Merritt Parkway in Westport, take Exit 42 and jog southeast to Route 136 (Easton Road). Travel northeast on Easton Road a few miles, turn right onto Treasure Road, left onto Gilbert, and right onto North Street. Park at the trailhead at the end of North Street. Follow the trail up into the

Brett Woods Nature Preserve. After you have reached the top and begin to descend, look for a side trail to your left. This side trail leads to numerous outcrops near the highest point of Brett Woods.

What you will see: Outcrops along the trail are Ordovician Ratlum Mountain Schist, a gray, medium-grained schist rich in biotite. Other minerals may be present, such as pyrite and garnet, which will stand out as blocky grains in the mica layers. You can see a big difference in the surface texture of schist, depending on how it is oriented. Glacial ice has smoothed the tops of some of the outcrops, while the broken edges of the foliation (mica layering) in schist make a rough surface. Because of the mica, schist tends to be somewhat plastic or pliable, so tectonic forces may produce many more folds than you may see in gneiss, even where they are adjacent.

Things to discuss: Medium-grained means you can see the individual grains with your eye, while fine-grained minerals may require a magnifying lens to make out and identify. How well can you see the grains of this rock?

Photo: Greg McHone

Stockbridge Marble at Kent Falls, Kent

Limestone and marble are among the best-known rock types of the eastern edge of Proto-North America, because they have been extensively quarried. The Connecticut capital building has a lot of marble from northwestern Connecticut. Rather than poke around a quarry, you can visit a "natural" outcrop of marble at Kent Falls State Park in northwestern Connecticut.

How to get there: Travel north on Route 7. The park is located approximately 4 ½ miles north of the village of Kent. There may be a parking fee on weekends and holidays in season.

What you will see: Kent Falls Brook descends from harder rock formations in the east and into Ordovician Stockbridge marble (named after the town in western Massachusetts). Slightly more resistant sections of marble created a series of falls, and potholes formed where cobbles of hard rocks were rotated by the water. Wherever streams flow down mountain sides, chasms, ravines, and valleys are rapidly formed.

All of these erosional features must have been made since the glaciers left this area about 16,000 years ago, but it is still easy to find bedrock surfaces that show hardly any post-glacial changes at all. Bedrock surfaces are efficiently shielded from erosion if covered by soil or glacial till, but where stripped bare we can expect at least a few inches of rock to be removed per thousand years. You can walk up a trail along the falls to see how the marble resists the water in some places better than others.

The marble is both smoothed and cut by the water, although its layers are tilted steeply to the east. You can also see folds and bends in the marble caused by the mountain-building events of the Paleozoic Era. Some marble is nearly white, but it is more common to see gray to brown patterns caused by iron and clay minerals, or other impurities. Fossils that might have been present would be extremely helpful for establishing rock ages, but they were generally destroyed by metamorphism. However, the Stockbridge Formation is well correlated with limestones farther west that contain Ordovician marine fossils (there were no land organisms at that early age).

Things to discuss: Marble is not very resistant to erosion by water. What might happen to this waterfall during the next million years or so?

The Newark Terrane 5

Photo: Greg McHone

The world was a very different place during the Mesozoic Era. We know from the geological record preserved in rocks formed during this time that high mountains ran up the middle of the region that would become Connecticut. Wide river plains spread out below across much of the landscape and supported natural communities populated by fantastic collections of fish, amphibians, reptiles, insects, plants—and dinosaurs.

The stories rocks of the Newark Terrane have to tell seem almost as if they were pulled from pulp science fiction—and yet there is no doubt they are real. It is a tale of a super-continent being ripped apart, of high mountains being eroded, of volcanic eruptions and of life and death. According to some scientists, the record of Mesozoic time in Connecticut rocks also reveals a time series when a global, environmental catastrophe occurred, causing mass extinction and forever altering the course of life on earth.

The Mesozoic Era (the Middle Ages in the history of life) began 250 million years ago, following a mass extinction at the end of the Permian Period that wiped out 90 percent of living species on earth. Life slowly rebounded and grew more diverse until there was another period of extinction, around the end of the Triassic Period and the beginning of the Jurassic. The end of the Mesozoic Era is marked by yet another extinction, one that brought an end to the age of dinosaurs some 66 million years ago.

Some of what is known of these events has been learned only recently by geologists working in the Hartford Basin. The basin has an astonishingly rich history of geological science going back more than two centuries to Benjamin Silliman and Edward Hitchcock, but the complex nature of the basin and its rocks continues to be revealed in new detail today.

The sedimentary rocks of the basin are of particular interest because they are known to preserve a unique and detailed record of more than 20 million years of Mesozoic time, beginning in the Late Triassic. The traprock ridges are also receiving renewed attention since geologists have come to discover they may hold clues to large-scale volcanic events thought to also have occurred during this period. As a result, the state's central lowlands figure prominently in scientific debates going on around the world today.

The global view

Two scenarios emerge as possible causes of the extinctions that occurred at the beginning, early-middle, and end of Mesozoic time. Among theories of how ecological disasters of these magnitudes could have occurred is one that suggests that asteroids called bolides may have struck the earth and exploded. Evidence that there was a bolide impact at the beginning of Mesozoic time was recently reported to have been found at Graphite Peak, Antarctica. Geologists analyzed rocks fragments from the site, rocks as old as the earth, and found they had chemical compositions considered unique to meteorites. They also reported finding shocked quartz,

*A mural at Dinosaur State Park depicts Dilophosaurus hunting along
a Mesozoic lakeshore like the one preserved at Dinosaur State Park.*
Photo: Brendan Hanrahan

a mineral formed under the great heat and pressure of an impact.
The evidence led them to conclude that a rock from space struck
the earth and exploded some 251 million years ago, near what is
now western Australia.

A meteor impact at the end of the Cretaceous Period and
the beginning of the Tertiary Period is now widely accepted as the
cause of the extinction that brought an end to the age of dinosaurs.
Known to geologists as the K/T boundary extinction (scientific
shorthand for the boundary between the Cretaceous and Tertiary
periods), the impact left a huge crater along Mexico's Yucatan
Peninsula.

Such impacts would have disastrous effects on life on earth.
A short time after the collision, dark clouds of smoke would have
risen into the upper atmosphere and been carried swiftly around
the globe. Clouds of smoke may have blotted out the sun and

plunged the world into cold and darkness. The delicate balance of ecosystems would have been suddenly disrupted. Many would cease to function. Lacking light for photosynthesis, plants might wither, leaving animals to starve.

There have been dramatic examples in recent years of just how quickly toxic clouds of gas and soot can spread when fueled by fire and heat. In 1980, the eruption of Mount St. Helens sent smoke and ash across several states and across the Pacific.

The eruption of Mount St. Helens in May, 1980. Photo by Austin Post, USGS/CVO/Glaciology Project

An explosion and fire at the Chernobyl nuclear power plant in 1986 sent clouds of radioactive particles across thousands of miles, reaching Europe and parts of Asia and North America in a few weeks. The remnants of both events are still easily detected in the soil in many of the affected areas.

Geologist Paul Olsen, of the Lamont-Doherty Earth Observatory of Columbia University, argues that yet another bolide impacted the earth to cause a mass extinction at the boundary between the Triassic Period and the Jurassic Period. Olsen is a world-renowned expert on the geology and paleontology of the rift basins of eastern North America. It is Olsen's opinion that the extinction at the Triassic-Jurassic boundary was most likely also the result of a bolide impact. He thought he found an impact site in a crater near Manicouagan, Quebec, Canada, but attempts to date the crater's age have been inconclusive.

Regardless, Olsen believes the Triassic-Jurassic extinction is significant because it gave early dinosaurs an opportunity to grow and expand, and set the stage for mammals (including humans) to later succeed them.

Another leading theory for a cause of extinctions involves volcanic events believed to have spread floods of lava across and through large portions of the earth's crust. Known as Large Igneous Provinces (LIPs), these floods mark times when geologic processes deep within the earth, in the section known as the mantle, pushed liquid rock up in such a way as to cause dramatic repercussions at the surface. It is possible these volcanic events may also have influenced the course of life.

Among of the most intriguing mysteries of Mesozoic geology is the fact that LIPs closely coincide with periods of extinction. The Siberian Traps, for example, were lavas that flowed at the time of the end-Permian extinction, at the beginning of the Mesozoic Era. A second flood, which is believed to have covered a large part of Connecticut and central Pangaea, has become known as the Central Atlantic Magmatic Province (CAMP). The CAMP lavas flowed practically right at the Triassic-Jurassic boundary. Paul Olsen is quick to acknowledge that they may well have played a role in the extinction he believes occurred at the boundary. The flood lavas known as the Deccan Traps flowed at the time of the extinction at the end of the Mesozoic Era, the K-T boundary.

Were the occurrences of the LIPs merely coincidental or

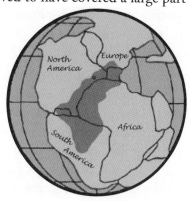

The Central Atlantic Magmatic Province. The dark gray indicates the area of Pangaea (including New England) some geologists believe was overspread by flood lavas beginning about 200 million years ago. Illustration by Greg McHone, after a map by Paul E. Olsen.

did volcanic events play a role in the extinctions? The fact is that no one is sure, but LIPs could have caused catastrophic change much the same way a bolide impact might have, by spewing large quantities of toxins high into the atmosphere.

The most likely scenario involves the injection into the upper atmosphere of large amounts of sulfur dioxide and carbon dioxide, causing drastic, if temporary climatic cooling, followed by heating. High lava fountains may have sent super hot gases and chemicals into the upper atmosphere, where they would quickly circle the globe. The result would be like a bolide impact and could produce global climatic change severe enough to cause massive extinctions.

The disturbing possibility is that a bolide strike was quickly followed by massive volcanism. It may even be that bolide impacts have contributed to or caused periods of volcanism. It could be that several of the greatest extinctions of the past coincided not only with lava floods, but also followed the impact of a large meteor— and that a combination of factors must be considered as causal, rather than any one in particular.

Connecticut in the Mesozoic

The rocks of the Hartford Basin preserved a considerable record of events in earth history and of living things that existed in the region during Mesozoic time. The basin's rocks reveal places where large lakes expanded and receded following cycles of wet and dry seasons, where lavas flowed into the lakes, where streams meandered to deposit great banks of gravel and where faults broke the earth's crust to create a geological pattern known as basin and range.

In a day or two of exploration, you can visit several places where the landscape, climate, and life of Connecticut in Mesozoic time is revealed. Most of the locations are in central Connecticut, but smaller sections of Mesozoic rocks are found in Southbury and Woodbury, in the Pomperaug Basin of southwestern Connecticut.

The Hartford Basin Terrane is essentially the central

Connecticut Valley, which runs though our state and north through
western Massachusetts. This chapter describes sites where Mesozoic
rocks and fossils can be found in East Hampton, Portland,
Rocky Hill, Manchester, East Granby, Middlefield, Meriden, and
Southbury. In preparation for making a visit, an overview of the
region of the state in Early Mesozoic time will help develop an
understanding of the significance of these locations.

The Newark Supergroup

The Hartford Basin, which forms the Connecticut Valley, is
just one of many Early Mesozoic basins that exist in eastern North
America between the high ranges of the Appalachian Mountains
and the North Atlantic Ocean crust. The famous Palisades,
across the Hudson River from New York City, is a basaltic cliff
on the eastern end of the Newark Basin of northern New Jersey

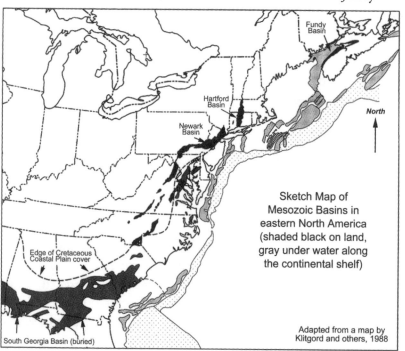

Fundy
Basin

Hartford
Basin

Newark
Basin

North

Edge of Cretaceous
Coastal Plain cover

Sketch Map of
Mesozoic Basins in
eastern North America
(shaded black on land,
gray under water along
the continental shelf)

South Georgia Basin (buried)

Adapted from a map by
Klitgord and others, 1988

and southeastern Pennsylvania. Other basins on land include the Gettysburg Basin of Pennsylvania, the Culpeper Basin of Virginia, and the Durham Basin of North Carolina. There are also large basins underwater along the continental shelf, including the Fundy Basin between Nova Scotia and New Brunswick. An extensive basin called the South Georgia Rift is covered by younger strata along coastal areas of the south. Because the sedimentary strata and structures of these basins are all similar to those first detailed in the Newark Basin, the group of basins that formed along what is now eastern North America are referred to collectively as the Newark supergroup.

Newark supergroup basins all contain Late Triassic sediments deposited before the Atlantic Ocean was stretched open, and when North America was still connected to Europe, Africa, and South America to form the super-continent of Pangaea.

Pangaea was too big to remain stable for long. By the middle of the Triassic Period, the earth around Connecticut began to move. Sometime around 225 million years ago, excessive heat in the mantle caused molten rock in the earth's core to flow in what are known as convection currents. Convection worked to thin portions of crust and to stretch it apart. Linear depressions called basins gradually began to form along the eastern side of the Appalachian Mountains. The earth began to slowly stretch apart where the Atlantic Ocean would separate Africa from North America in a process geologists refer to as continental rifting.

The Appalachian Highlands formed a western boundary for the basins. The root structures of the mountains extended deep into the crust, and restricted the more shallow faults that defined the new basins. To the east, the Hartford Basin was bounded also by the Bronson Hills, then tall mountains. Pangaea stretched to the northwest and southeast, but the terrane defined by the Bronson Hills held the basin to its north-south orientation. Streams swept masses of boulders and smaller stones down out of the hills and spread wide fans of conglomerate across the valley's river and lake sediments. As the surface relief increased east and west of the basin, sediments in those areas were gradually eroded to where the basin today is considerably smaller than it was in the Triassic Period.

As Pangaea stretched, periods of volcanic activity erupted. The surface lava flows (known as basalts) of the Hartford Basin, and the underground systems that fed them (known as dolerite dikes and sills) were widespread across central Pangaea. The area of the CAMP extended from modern Brazil, northeastward across western Africa, Iberia, and northwestern France, and from Africa westward through eastern and southern North America as far as Texas and the Gulf of Mexico. Soon after this Early Jurassic volcanism, North America broke away from Africa. Many Connecticut basalts may in fact be similar to the crust underlying the oldest areas of Atlantic Ocean seafloor.

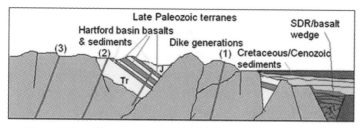

Cross section of the northeastern USA rift zone (southern New England)

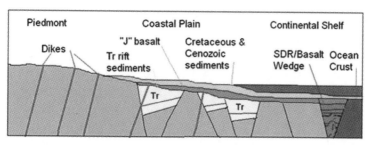

Cross section of the southeastern USA rift zone (South Georgia rift)

The Early Jurassic continental rifting of Pangaea included activity of Mesozoic basin faults, especially movements of the great border faults alongside every basin. The continent along and east of the Appalachians was uplifted at the same time that the basins tilted and dropped down and along border faults, which allowed most of the Mesozoic strata between basins to be eroded away. Gradually through the Jurassic Period, tectonic and erosional events made the basins into their present geographic patterns.

The basin strata may be tilted and faulted, but they are not nearly so deformed as the older Paleozoic metamorphic rocks beneath them, often called "basement rocks." The metamorphic basement contact with the oldest Triassic sandstones of Connecticut is poorly exposed, except at one place along Roaring Brook in Southington. There you can clearly see the major unconformity between the two different terranes, which represents a time gap of about 180 million years. Unfortunately, the "Great Unconformity" is on private property and is not accessible to the public.

A generally eastward tilt of 10 to 20 degrees has affected all rocks in the Hartford and Pomperaug Basins, except where local faults have rotated blocks of strata into different attitudes or angles. The Law of Initial Horizontality states that sedimentary layers (and to some extent lava flows) always form close to level or horizontal, which means the tilting occurred sometime after Early Jurassic sedimentation ended. Because the strata are more than four miles thick in total, there must have been a lot of faulting to juxtapose the strata against the same rock formations that underlie the basin, assuming that the strata formerly continued to the east. However, most geologists believe that faulting was also occurring at the same time that sediments were being deposited in the basin, so it is not clear where and which strata were originally present across the border fault.

The tilt of the Mesozoic strata is especially evident along highway cuts. The most famous of these is along Route 9 in East

Berlin between I-91 and Route 15 (the Berlin Turnpike). Easily seen from a car or bus, it is neither safe nor legal to stop at these places without first obtaining a special permit from the State Highway Department. Even with a permit, no one should attempt to stop at these places without an expert to guide them.

Mesozoic sedimentary rocks

Most of the Mesozoic rocks of the Connecticut Valley are made from sand and gravel deposited in streams and rivers, also called fluvial deposits. Between 225 million and 195 million years ago, sand, gravel, and mud carried by streams were deposited in layers in the Hartford Basin, gradually reaching thousands of feet in thickness. Average rates of deposition were very high, as much as one to two feet of sediment per thousand years.

Water moving through the ground (groundwater) carried minerals such as iron, calcium (lime), and silica, which coated the sand grains and eventually cemented them together to form sedimentary rocks. Sandstone made with quartz, feldspar, and clay is also called arkose. Brownstone is the variety of arkose perhaps most familiar to people in Connecticut. Its rust brown color comes from the iron ore that formed its cement. In some places, the iron can make these rocks appear almost red or purple. Because of the rich color and ease by which it can be quarried and cut into building stones, brownstone has been a very popular building material throughout the state's history.

A general term for layers of rocks that collect on the earth's surface is strata. The agent that deposits most strata is water, but wind, volcanoes, or even landslides are also possible. Geologists who study these rocks are stratigraphers, and their field of study is called stratigraphy. Groups of strata that are similar, formed together, and can be traced across the land, are organized into formations with names such as the East Berlin Formation. A formation is usually named after a place where it is especially well exposed or studied.

Sandstone conglomerate
Photo: Greg McHone

Sandstones in Connecticut often have alternating sand and gravel layers that overlap, indicating deposition by streams and rivers that wind and meander as they flow. During Late Triassic time, rivers flowed from the northeast to southwest across the state in wide, flat plains, not at all like the steep valley rivers we have today. Some areas of very coarse gravel and cobbles were brought by streams with a lot of energy, indicating steeper slopes. Hills and low mountains that rose up out of the plains would account for the areas of faster stream flow and coarser sediment.

Sediments carried and deposited by rivers and streams are known as fluvial deposits. The rust red color of local brownstones is an indication they were exposed to air on land or in shallow streams and lake shores.

Other layers, especially those found between lava flows in Early Jurassic time, are made of silt and clay that could only have settled out of quiet bodies of water such as lakes or sluggish rivers. The term for lake-deposited strata is lacustrine. When these materials become cemented, they can be called mudstone or siltstone, and if abundant clay causes the rock to easily cleave or break along flat planes it is usually called shale. Some shale layers are very dark gray due to organic materials that were preserved in the mud by chemical conditions

Layers of shale
Photo: Greg McHone

that lacked oxygen, as may be found at the bottom of deep lakes. These "reducing" conditions can also cause sulfide minerals such as pyrite to form, which corrode to rusty surface stains after the rocks become exposed to air.

Still other sedimentary layers are much lighter brown or tan in color. This color is an indication the sediments were deposited during dryer climatic cycles.

Volcanic lavas, dikes, and sills

The traprock ridges of the Hartford Basin were formed by three huge, volcanic lava flows known to have occurred in the valley about 200 million years ago. They are not very high, perhaps a few hundred feet, but their rugged western faces and tall cliffs of eroded, columnar basalt give these central ridges their appearances.

To the east is a ridge system of basalt strata that includes the Hanging Hills of Meriden, Totoket and Higby Mountains of the Metacomet Ridge, which reaches up to the Holyoke Range of Massachusetts.

To the west are the visible portions of large, intrusive magmatic sheets (known as sills) formed by the first basaltic magma. All are made of dolerite, the intrusive equivalent of basalt. The sills solidified thousands of feet beneath the surface and have since become exposed due to tilting and the erosion of much of the surrounding sandstone. This smaller, western system in Connecticut includes West Rock, Sleeping Giant and the Barndoor Hills.

The sills and flows are hundreds of feet thick, but not unlike thinner basaltic flows erupting today in Iceland, Hawaii and elsewhere. They are exactly like Early Jurassic basaltic flows and dolerite sills found in other Mesozoic basins along eastern North America. The basalts all erupted during the same interval of 580,000 years soon after the start of the Jurassic Period, they all flowed on sediments of the same age in different basins and the composition of the basaltic magmas are similar in all basins. These correlations among the basin basalts are very close, and support the

Basalt at East Peak, Meriden
Photo: Greg McHone

hypothesis that they all were part of the same lava flow that spilled across 1,000 miles of the rift basin zone.

An essential question has been how lava could flow over such great distances without cooling and solidifying. Until the 1980s, most geologists assumed the volcanoes had formed separately in eastern North America and produced separate magmas. The lack of evidence of volcanoes was attributed to erosion and burial beneath sedimentary strata.

A new model has gained acceptance only in recent decades. Geologists have collected evidence that great flood basalts found in Connecticut flowed instead out of fissures, cracks going down 40 miles or more through the crust and into the mantle, where rocks rich in iron, calcium and magnesium are melted into magma.

Fissures large enough to connect eastern North America have recently been recognized in the form of dolerite dikes. These dikes were formed from magma that froze within fissures after they stopped feeding surface flows. While physical connections with the surface have been eroded or buried, comparisons can now be made between the chemical compositions, magnetic properties, and ages of dikes with surface basalts. Recent evidence provides an excellent match between systems of dikes and the lava flows.

The first eruption of about 201 million years ago flowed from a fissure across a large portion of what is now northeastern North America. This dike has since been documented from Branford, Connecticut to Massachusetts, into southeastern New Hampshire and coastal Maine, all the way to New Brunswick, Canada. These dikes are today among the largest known on earth. The Higganum Dike can be clearly seen in Connecticut and takes its name from that village of the same name on Route 154 in Haddam.

The Talcott Basalt flowed from this first fissure in central Connecticut. It has been shown to be the same magma found in other rift basins of eastern North America. Because the fissure system is continuous in those areas, it is likely that Talcott-type lavas covered much of the land in its path and not just where it remains today in basins.

Where wetlands were overrun by the 2500-degree lava, steam boiled upward into the flows, breaking the hot basalt into zones known as hornitos. In some places, plants and small tree trunks were preserved as baked, coal-like forms in the mud under the lava. Where the lava flowed into lakes, blobs of glassy basalt with shapes of overlapping pillows extruded from cracks along the front of the underwater flow.

The greatest of the three volcanic formations is the middle layer known as the Holyoke Basalt. This layer is up to 500 feet thick in places and forms most of the traprock ridges of the Hartford Basin. The third and uppermost flow is the Hampden

The Buttress Dike
Photo: Greg McHone

Basalt, which is up to 300 feet thick. Each basalt formation covers hundreds of square miles today, but may actually have flowed over much greater areas. Fissure dikes or formations that fed lava to the surface run across all of southern New England and perhaps beyond. They are of truly huge dimensions and may have actually been created by several episodes of volcanic eruptions from their fissures.

After the first eruption of about 201 million years ago, the next two volcanic events were spaced about 300 thousand years apart. Each basalt formation has its own dike source: The Fairhaven-Higganum Dike System fed the Talcott Basalt, the Buttress Dike System fed the Holyoke Basalt, and the Bridgeport Dike System fed the Hampden Basalt. These dike systems run in a southwesterly direction across Connecticut and Massachusetts, and there are other very large dikes in Maine and New Brunswick that may be extensions of the Higganum and Buttress Dikes. If so, these intrusions are true giants of the earth, and their effects must have been catastrophic.

Another traprock ridge system is present in the western and southern areas of the basin, including East Rock, West Rock, Sleeping Giant, and the Barndoor Hills. These ridges are technically diabase, not basalt, because their magma cooled as large intrusive sheets (sills), deep within the strata of the New Haven Arkose. Due to the effects of tilting and erosion which occurred later, the sills have come to look like the extrusive basaltic lava formations tilted eastward. All of the intrusive sills were derived from the same Fairhaven-Higganum Dike System as the Talcott Basalt and flowed during the same time. Apparently the New Haven Arkose was still soft and loose when the dike reached it, encouraging the magma to spread outward along the layers as well as upward to the surface. However, the immense heat from all that magma caused hot mineralized water to circulate, which rapidly cemented the arkose into solid stone. One hypothesis (not yet even a theory) is that the New Haven Arkose was made so solid and brittle by the first flow that the second and third volcanic flows, the Holyoke and Hampden Basalts, were prevented from flowing anywhere but straight to the surface.

The Hartford Basin "layer cake"

By the late 1970s, geologists came to understand the stratigraphy of the Hartford Basin, including its formation thicknesses, ages, and characteristics. Working up from the bottom, or from oldest to youngest, the New Haven Arkose is the oldest sedimentary formation of the Hartford Basin. This formation is about two miles thick and of Late Triassic age.

Above the New Haven Arkose is the Talcott Basalt, a 300-foot-thick section of Early Jurassic volcanic rock and the first of a group of three basalts separated by layers of mudstone. Above the Talcott

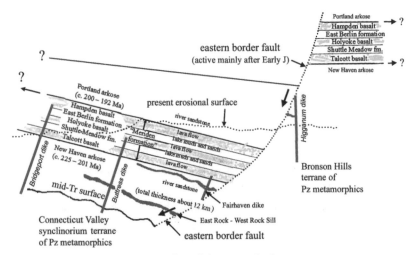

Stratigraphy of the Hartford Basin
(Theoretical schematic proposed by Greg McHone)

Basalt is the Shuttle Meadow Shale. Next is the second lava flow, the Holyoke Basalt. On top of it is the East Berlin Siltstone. The last lava flow is known as Hampden Basalt. At the top is the Early Jurassic Portland Arkose, a layer that may originally have been as much as two miles thick.

There are also wide areas of coarse fluvial conglomerates within the Early Jurassic formations near the eastern border fault, which is evidence for fault-related episodes of higher topographic relief. Because of the eastern tilt along with general regional uplift, all of the formations have had their western ends eroded away, and much of the upper part of the Portland Arkose is gone as well.

The Triassic-Jurassic boundary

Since Paul Olsen has advanced his theory for a Triassic-Jurassic extinction, there has been a lot of interest in the exact rock level where the Triassic Period ends and the Jurassic Period starts. This boundary line is in the New Haven Arkose about five to ten feet below the contact with the Talcott Basalt. It is marked by abundant fossil ferns and fern spores.

About two-thirds of the plant and animal fossils found below this level (the level of the Triassic-Jurassic boundary) are not found above it. In the field of biostratigraphy (the study of the evolution of life as organized by geological formations), places where fossils are found to suddenly disappear from strata may provide evidence of periods of extinction. Paul Olsen interprets the changes in fossil forms found below and above Hartford Basin strata marking the Triassic-Jurassic boundary as evidence of a great mass extinction, one that enabled dinosaurs to rapidly expand, grow larger and come to dominate life on earth.

Environments, plants, and animals

Connecticut was located near the center of Pangaea and close to the equator during the Early Mesozoic. Not only was it a hot, dry place, it was in the middle of a hot world. There were no Arctic or Antarctic ice caps, and conifer forests stretched across Pangaea to the south and north poles. Seasonal changes were less extreme, although it must have dropped below freezing during the long dark winter months in the polar regions. The climate overall did not become more temperate until late in the Cretaceous Period.

Climatic conditions were in part a result of the very high concentrations of carbon dioxide (CO_2) that existed in the atmosphere. During the Mesozoic Era, CO_2 in the air was as high as 1000 ppmv (parts per million by volume) or about three times as concentrated as it is today at about 380 ppmv (and rising). CO_2 is a "greenhouse gas;" it traps solar heat and holds it in the atmosphere. Without CO_2 in the air we would all freeze, but we would bake if there was too much.

Mesozoic Connecticut was dry most of the year, but there was a rainy summer season when a monsoon would arrive and fill the lakes and rivers, at least temporarily. Connecticut's landscape might have looked like parts of today's Mexico or East Africa, or perhaps southern Arizona. Tall conifer trees dominated the forests where there was enough rainfall, accompanied by ginkgo trees and

The arid landscape at Elephant Head, Arizona
Photo: Nancy McHone

tropical species such as tree ferns and cycads. Wetland marshes and lakesides harbored abundant ferns, reeds and horsetails. In Early Mesozoic time there were no flowering (angiosperm) families of plants, which evolved later and replaced many older (gymnosperm) types of plants during the Late Mesozoic. In Connecticut, there were no grasses or wildflowers. There was not the great variety of trees and shrubs we see today.

Without flowers, there could be no butterflies or bees as we know them today. Birds had not yet evolved, but flying reptiles (pterosaurs) and large insects such as dragonflies flew among the trees. In the rivers and lakes were abundant fish with bony heads and strong fins, similar to coelacanths (the "living fossils" that still exist in the South Pacific), along with turtles and amphibians. A wide variety of invertebrates such as worms, shrimp, and shellfish were also present.

On land the reptiles were dominant. Late in the Triassic Period there were long-legged crocodiles, fierce phytosaurs (which looked like crocodiles), as well as early mammals. Dinosaurs appeared in the Late Triassic, but were small and relatively few. Paul Olsen believes they became larger and more diverse only after other, more primitive reptiles went extinct at the Triassic-Jurassic boundary. During the Jurassic and Cretaceous Periods, dinosaurs expanded into the more familiar forms and came to dominate life on earth. The sedimentary rocks of the Hartford Basin record the time following the extinction when dinosaurs first began to expand. Olsen has presented evidence to show that meat-eating theropod dinosaurs suddenly appeared that were much larger than any known from the Late Triassic. Plant-eating dinosaurs also grew larger and more abundant during this time.

Dinosaur tracks in Connecticut

A wide variety of tracks made by dinosaurs, reptiles, insects, arthropods, and other invertebrates have been studied in the Mesozoic rocks of the Connecticut Valley, and in similar basins elsewhere in eastern North America. The science of tracks and other trace fossils is ichnology (or stony footprints) and the fossil footprints are known as ichnites. Because many different dinosaur species could make very similar footprints, and no dinosaur skeletons have been found with the tracks, we cannot assign particular animal species to particular footprints. Instead, footprints have their own system of names that are defined separately from any animal species. Hundreds of footprint "ichnospecies" have been identified since Edward Hitchcock published the first study of fossil footprints in 1836, but scientists including Paul Olsen have argued more recently there should be far fewer. Paul has shown how differences between individuals of different ages and sexes, as well as differences in mud surfaces and preservation, could easily account for previously perceived species variations.

Other, nondinosaur tracks are also common in the Connecticut Valley. Some were made by a variety of large and small four-footed reptiles similar to *Stegomus* (an armored herbivore), *Terrestrisuchus* (an early long-legged crocodile), and *Rutiodon* (which looked like modern crocodiles). A variety of crustaceans and insects also left their own tiny trace fossils.

Visiting Mesozoic sites in Connecticut

Natural Mesozoic features are abundant in Connecticut, but many are in dangerous locations along highways, where it is illegal to stop or pull over, or on restricted private locations. Many of the locations described in this chapter are accessed more safely, although some are on private property. When you visit sites, please be respectful and do not climb on rocks, leave litter or take anything away. There are other state and town parks that show Mesozoic features, especially great basalt (traprock) exposures at East Rock Park and West Rock Park near New Haven, Sleeping Giant State Park in Hamden, Wadsworth Falls State Park in Middlefield, and Talcott Mountain State Park in Simsbury. After you learn a little more about these rocks, visits to any of these locations will be rewarding.

Higganum Dike at Hurd State Park

The huge fissure eruption that produced the Talcott Basalt around 201 million years ago flowed from a crack through the earth's crust, which is still visible because of the frozen basaltic magma in it. Such magmatic cracks are called dikes, because in some places they stand above the surface like a dike or wall that could hold back water. The Higganum Dike is named from its exposure on Route 154 in the village of that name, but there we have no safe place to park or walk to look at it. Another well-known location is at Exit 9 off Route 9, but there is no access to this location either. Do not attempt to stop or walk near these locations. A much better site, where the dike can be viewed legally and safely, is easily accessible at Hurd State Park.

How to get there: Hurd State Park is located just off Route 155 near Haddam Neck in East Hampton. From East Haddam, drive northward on Route 155 until you see the sign for the park, or from Portland drive eastward on Route 66 to Cobalt and turn south onto Route 155, through Middle Haddam, and onto the Haddam Neck Road to the park. Drive through the gate into the park, and after about a third of a mile turn right to go up a hill to the picnic pavilion. Park and walk across the field opposite the pavilion to a low wooded hill, which is an outcrop of the dike.

What you can see: The blocks of diabase (basalt that cooled inside the earth) have been broken up by winter frost in fractured columns in the dike.

Where it is more resistant than the surrounding metamorphic rocks, as at this location, the dike forms a low hill or ridge. In other places it is covered and can only be located by its higher magnetism relative to surrounding rocks. This same dike system has been traced from Long Island Sound northeast through Connecticut, New Hampshire, the coast of Maine, and into New Brunswick—

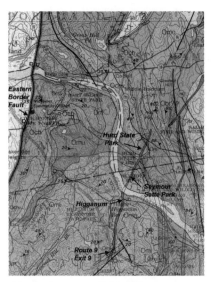

Geologic map of the Higganum Dike at Hurd State Park and other locations in Haddam.
Courtesy, Connecticut Department of Environmental Protection.

almost 400 miles! Its eruption and flood basalt was perhaps the single largest volcanic event ever to occur on earth.

In 1997, a sample of the Higganum Dike at Hurd State Park was analyzed by the argon gas radiometric method and determined to be about 201 million years old. Lavas from this fissure must be the same age. If you will be visiting one of the basalt stops described later in this chapter, you might compare basaltic lava that flowed onto the surface with this equivalent diabase that cooled deep inside the earth. There has been about five miles of erosion of the earth's surface after the dike formed, so the magma here was still far below the fissure volcano.

Notice that the rock is fine-grained, but mineral grains are just visible to the eye, including tiny pale crystals of plagioclase feldspar that stand out on weathered surfaces. For contrast, walk back across the field and down the trail behind the Pavilion to see pegmatite in an abandoned quarry. This very coarse igneous rock mainly contains feldspar, but it cooled at an even deeper level about 280 million years ago. Pegmatite is less dense (lighter in weight for a similar size) than diabase, which has a lot more iron, magnesium and calcium in it.

Things To Discuss: Now that you see what a diabase dike looks like, how hard might it be for a geologist to locate and map it across the state?   Does this rock look like something useful for buildings such as walls and foundations? How can you tell that it contains iron minerals?

The rest of the locations described in this chapter concern Mesozoic rocks in the Hartford Basin (Connecticut Valley) and the Pomperaug Basin of southwestern Connecticut. To see some of these features, head north on Route 155 and turn west onto Route 66 toward Portland.

Columns of Higganum Dike at Exit 9 of Route 9, Haddam.
Photo: Greg McHone.

Photo: Greg McHone.

Brownstone (Arkose) and Conglomerate, Route 66

The Portland Formation of Mesozoic sediments has a lot of the sandstone called brownstone, which was quarried in Portland and other locations in the Connecticut Valley. But there were other types of sedimentary materials deposited as well, some of which we will see at this stop.

How to get there: From Main Street in Portland, turn southeast on Route 66 and travel about two miles, or if you are headed west on Route 66, go through Cobalt toward Portland and past the intersection with Route 17, and turn into the "Prehistoric Golf" shopping plaza. Park toward the left. Walk up the paved area to the brownstone behind the hardware store. This is private property, so please do not climb or pull down rocks from this fine exposure. If you are with a group, let the manager of the hardware store know that you wish to look at their rock cut. In the summer you can play a round of miniature golf among the life-sized dinosaurs on the hillside -- it seems like an appropriate location!

What You Can See: Here we are near the top of the Portland Formation of Early Jurassic age, perhaps about 192 million years old at this level of the valley. Much of the Portland Formation rocks are arkose, which is sandstone that contains feldspars as well as quartz grains, often with some silt and clay mixed in. There were plenty of these minerals being eroded from older rocks of New England at this time. Some of the mineral grains were carried in by rivers from the north and east, while other portions of these sediments were derived from nearby hills. The great fault that forms the eastern border of the Connecticut Valley became more active around this time, and it is nearby; the eastern uplands start not far to the east, about where Route 66 heads uphill.

The rocks on the left side of the cut are very fine grained, and they break or "cleave" along flat surfaces, a characteristic of a type of mudstone called shale. Shaley cleavage in mudstone is caused by clay in the sediment. Individual grains of silt and clay are too small to see by eye (try a magnifying hand lens if you have one), but you can rub them off as a powder with your fingers. Mud this fine grained might have been deposited in a lake, where it could slowly settle out of the quiet water. Where the muddy bottom became exposed along shorelines, dinosaurs, reptiles, and insects have left their footprints in rocks like these.

Walk along the cut to the right, and notice how the shaley cleavage is replaced by lenses and layers of much coarser materials, which change the character of the rock (but not the purplish-brown color, which is caused by iron added to the sediments by groundwater). The sand, gravel, and cobbles are all large enough to see without magnification, even after they are cemented together. Here a fast stream dumped sand and large stones into the lake covering the mud, which must have flowed down from nearby hills. Mixtures of rocks like this are called conglomerates. About this time the nearby eastern border fault probably moved and dropped the valley downwards, so that streams from the eastern hills became steeper, faster, and able to carry coarser material into the valley. Some of the fault activity could have been sudden, causing big earthquakes, but these movements could also have been slow and steady over hundreds of thousands of years.

Things to discuss: Pick up small samples of shale and conglomerate. Can you identify some of the individual mineral grains? Where have you seen materials like these before they became rocks? Can you imagine the Mesozoic lakes and streams that existed right where you are standing?

Portland Brownstone Quarries

The largest brownstone quarries are along the Connecticut River in Portland. After being abandoned for 60 years, the quarry area is now owned by the City of Portland and is being redeveloped into an historic park.

How to get there: To go to the brownstone quarries from the shopping plaza, head north into Portland on Route 66, turn left on Main Street and immediately right (north) next to the gas station, onto Silver Street. Proceed straight to the bottom of the hill and turn right. Park along the retaining wall next to the larger water-filled quarry.

What you can see: Amazingly, the top of the hill behind the quarry reached over your head to the river behind you when the first quarry operations started in the mid-1600s. Another large water-filled pit is off the road to the south (right). When freshly cut, the stone is relatively soft and can be carved, but then it dries out or "cures" to a harder, more brittle stone. All of this rock was removed to make buildings, monuments, and bridges, especially in the northeastern states and New York City, where "brownstone rows" built in the late nineteenth century are still popular residences. The quarries were abandoned after being flooded in the hurricane of 1938, but

demand for the stone had declined anyway. The water level in the quarries is close to that of the river, where boats used to dock to take on stone that was transported up and down the coast. Today, a smaller operation behind the trees to the north of the old quarry once again cuts brownstone, which is enjoying renewed demand for restoring old buildings as well as for new construction.

Dinosaur footprints in the quarry were so common that many were ignored and destroyed, but some examples were saved for museums and private collections. If you have time, visit the science library at Wesleyan University to see footprints from this quarry mounted on the walls of the main floor. The stone is typical Portland Arkose, as you can see in the blocks along the overlook walls. It was split into thick slabs along bedding layers created in the Jurassic river bottoms where the gravel was deposited, and the slabs were then cut into blocks with huge saws. Another helpful characteristic of this area is the nearly horizontal orientation of the bedding planes, unlike the more common 10 to 15-degree eastward tilt. The quarries must be part of a large block of the upper crust that was separated by faults from other blocks that became tilted.

Things to discuss: Do you like the texture and dark reddish-brown color of brownstone, and can you see why it might come into or out of style? Are there some brownstone buildings near where you live? Why would this and other stones be especially desired for buildings in cities (think fires)?

If you wish to continue to Dinosaur State Park, go back to Main Street (Route 66) and across the Arrigoni Bridge over the Connecticut River, and turn onto Route 9 North, and then I-91 North.

Photo: Greg McHone.

Dinosaur State Park, Rocky Hill

Dinosaur footprints are prized by collectors, and many have been illegally broken or cut out of rock exposures in public and private lands, which destroys most of their scientific value. It is much better to visit one of the parks and museums in Connecticut, where you can see tracks properly identified.

How to get there: Dinosaur State Park, in Rocky Hill, is easily reached from I-91 south or north. Take Exit 23 and turn right if your were headed north, and left if you were headed south, to go east on West Street. After about 1 mile or so, Dinosaur State Park is on the right (south) side of the road. The park is open Tuesdays through Sundays, and there is an entrance fee into the exhibit building, although the grounds and parking are free.

What you can see: This wonderful park has a large dome-shaped building that covers the trackway as well as housing many exhibits and a gift shop. Outside is a widely praised arboretum of conifer trees and plants similar to those that grew in the Mesozoic Era, such as gingkos, Monkey Puzzle Trees, magnolias, and others. There is a nature trail through a bog, and a special area where you can make your own plaster cast of a real *Eubrontes* dinosaur track.

The following history of Dinosaur State Park is excerpted from a 1985 field guide by John Rodgers and Brian Skinner, professors of geology at Yale University:

"In 1966, the State Highway Department chose this site for a central Highway Department Research Laboratory, close to but not on Interstate 1-91 near the geographic center of the state. One Friday afternoon in August, one of the bull-dozer operators, Mr. Ed McCarthy, turned up flat slabs of sandstone on which he recognized some large dinosaur footprints (such prints have of course been well known in the Connecticut Valley for 150 years)… The news also reached Ms. Jane Cheney, Director of the Children's Museum in Hartford, who went directly to Governor John Dempsey (about to stand for re-election) and persuaded him that the find was exceptional and should be preserved…. Later the Governor declared the locality "The Dinosaur State Park."

A news item concerning the dinosaur trackway appeared on the front page of the Hartford Courant for twelve straight days. Clearing continued for several weeks, until a single surface of sandstone displayed over two thousand tracks. Testing elsewhere on the property showed that the layer with the tracks was even more extensive; moreover it is only one of five layers within about 2 meters of rock that display tracks.

By this time, it was thought that enough had been uncovered to make a spectacular display, and the work was stopped; the main concern after that was to preserve the tracks against the approaching winter's freezes and thaws. The tracks were therefore covered up, and, except for one or two brief spells, the main discovery site has not been uncovered since… the original plan to build a larger museum over the main, original discovery has never been carried through. In any case, the Park was duly dedicated in 1967 by Governor Dempsey; honor was paid to Mr. McCarthy, the original finder of the tracks; and the Rocky Hill High School Band played a new piece of music called "Dinosaur," written for the occasion by its Director.

The Highway Department, deprived of their original site, had to recommence operations about a mile farther east, and rumor has it that the bulldozer operators were given strict orders to stop for nothing."

Start your tour by examining the time-line walk, which is described in a plaque near the parking lot. The amount of time recorded by fossils is only a small portion of the "deep time" history of our earth. It is clear that dinosaurs did not roam "when the earth was young," as is sometimes claimed, but rather after earth's history was well advanced, and it is humbling to see how humans have

The dome covering the dinosaur footprint trackway at the park.
Photo: Brendan Hanrahan

only been around for such a tiny and recent time! You can also understand that people and dinosaurs lived at very different times, unlike what you see in many movies and cartoons.

Also outside you can see a low exposure of the siltstones that are present in the park, part of the East Berlin formation that lies between the Holyoke and Hampden Basalt Flows. There is a plaque describing the field where hundreds of tracks had to be covered over, as mentioned above by Rodgers and Skinner. You can also tour the arboretum of conifers, and a nature trail that takes you through a wetland to the east.

Inside the exhibit building, walk slowly in a counterclockwise direction to see the history of the park and information about dinosaur tracks found in Connecticut. There is an excellent recreation of a road cut of the East Berlin Formation along Route 9 where it crosses the Berlin Turnpike, and you can learn a lot about the sediments and lava flows. Also, there is information and a life-sized model of *Coelophysis*, the Late Triassic dinosaur that might have made the smaller tracks called *Grallator*.

Around the corner are wonderful life-sized dioramas of plants and animals that lived here 200 million years ago, across from the actual *Eubrontes* tracks formed in a muddy lakeside surface that is now siltstone. Several large animals must have walked back and forth over this mud flat, and they may have been the dinosaur we know as *Dilophosaurus*, which lived right at this time. There he is looming next to the railing—see those sharp teeth and the fierce look in his eyes! At this time *Dilophosaurus* was king of the land, since larger carnivores such as *Allosaurus* and *Tyrannosaurus* had not yet evolved. Along the same side of the exhibit you can see other animals such as giant dragonflies and roaches, *Dimorphodon* pterosaurs (not dinosaurs but a type of flying reptile), and farther along, two herbivore dinosaurs: a larger animal called *Anchisaurus* (an ancestor to the huge sauropods such as *Apatosaurus*) and a smaller, birdlike dinosaur called *Lesothosaurus*. Some of the trees

*The casting area at Dinosaur State Park, where
visitors can make plaster casts of dinosaur footprints.*
Photo: Brendan Hanrahan

and plants look tropical but are otherwise familiar, including
horsetail reeds, ferns, conifer trees, and bushy palmetto-like trees
called cycadeoids. There were no grasses or flowering plants
(angiosperms) in existence at this time.

On the wall is a large mural by William Sillin called *In the
Late Triassic*, which represents this area a few hundred thousand
years before the *Eubrontes* tracks were made. The painting is a
view across central Connecticut before the first volcanic eruption
covered it with Talcott lava. The meandering river is just the sort
of place for brownstones to form. The small theropod dinosaur
is *Coelophysis*, and we again see two *Lesothosaurus* plant-eaters on
the right side. Out in front of the mural, notice the dangerous
Rutiodon (a phytosaur reptile, not a dinosaur or a crocodile, despite

the resemblance), which competed with dinosaurs in the Triassic Period. Luckily for dinosaurs, other large land reptiles disappeared by the start of the Jurassic Period. The large flat amphibian with teeth, called *Metoposaurus*, probably lived under water.

Things to discuss: The state's mission for this museum is mainly to provide educational programs for school children, which is certainly worthwhile. Many people already believe that rocks and dinosaurs are fine for kids but not very interesting or important for adult society. Does this concern you, and if so, how could the museum and park be used to better promote geology and earth history among the general public of Connecticut?

Photo: Greg McHone.

Arkose and faults at Buckland Hills

If you happen to shop in the Buckland Hills Mall area of Manchester, you cannot help but notice the red arkose, or coarse sandstones of the Portland Formation cut by several roads in and around the I-84 Exit. The mall itself was built near the old Wolcott Quarry, which is famous for being one of the very few places in New England where actual dinosaur bones have been found. You probably will not find any bones, but the stream gravels that dinosaurs walked along and faults that caused Mesozoic earthquakes are both well exposed. Plus, those in your family who like to shop will appreciate this location.

How to get there: If you are headed East on I-84, take Exit 62 onto the parallel road to a light, turn left onto Buckland Street and right at the light onto Buckland Hills Drive (between Lowes and Home Depot), then up the hill and left into the Target store parking lot. If headed west on I-84, take Exit 60 and turn right onto Buckland Street, then follow the above directions. Park around the right (northeast) end of the store near the rock cut.

What you can see: This is a particularly nice excavation into Early Jurassic Portland Arkose. Do not go up to the rock face, as you can see everything well from the edge of the pavement, where it is safe. Remember that this is private property. Look, but don't touch.

In the rock face you can clearly see overlapping lenses of coarser and finer gravel, which were formed by point bars inside meandering stream channels. The white quartz pebbles were eroded from older metamorphic rocks, probably not far to the east of this location. Some cobbles of schist and gneiss are also present. The deep red color is from iron oxide that precipitated from groundwater, eventually helping to cement the gravel into coarse sandstone. A fault slices across the face to offset the channels by a foot or so. The side to the right is over the left side and has slid downwards, which classifies this as a "gravity" or normal fault. When this motion occurred is not known, but a reasonable guess is that it was related to the larger movements of the great eastern border fault of the Hartford Basin, which was mainly active long after these rocks were deposited and solidified.

Things to Discuss: Compare these gravels to some from modern streams that you have seen. Was the river that deposited them deep and slow like the Connecticut River, or was it relatively shallow and fast moving? Why are bones somewhat surprising to find in this kind of sediment?

Photo: Greg McHone.

Tariffville Gorge, East Granby

To the north is a place popular for kayaking and fishing, but it is the exposures of Mesozoic rocks geology fans come here to see. Tariffville is an old mill village in Simsbury on the west side of the Farmington River—which has cut a gorge through the basalts and intermediate sediments.

How to get there: From I-91 North take Exit 38 onto Route 75, and then the Day Hill Road ramp. Follow Route 187 north, stay left where it merges with Route 189, and after you cross the bridge over the river on Route 187, turn right onto Spoonville Road, and immediately right again. The road goes beneath the bridge and follows the bank of the Farmington River upstream. Park at the cul-de-sac at the end of the road.

What you can see: The road cut and river bank exposures provide a nearly complete cross section through the three basalt and two intermediate formations, but poison ivy has covered some of the rocks, so beware! From the end of the road, descend directly to the Farmington River, not down the easy path used by the kayakers. As you climb down, you will cross outcrops of Shuttle Meadow Formation above Talcott Basalt.

A more complete section through this basalt is exposed across the river where this road once continued—the old bridge here was washed away some years ago. In the few feet exposed on our side of the river, typical pillow and gas-bubble-rich forms of the Talcott are visible. The Shuttle Meadow Formation consists of red mudstones and siltstones with a very pitted, weathered surface. The pits are caused by the dissolution of calcite, which filled spaces left behind by still earlier minerals.

Walk downstream along the river for a short distance. Across the river are excellent exposures of Shuttle Meadow siltstones, and if you can get far enough (in times of low water) the Holyoke Basalt appears as large ledges that stick out into the river. Note how the southeasterly tilt of the layers allows you to walk "up section" or from older into younger strata as you go down river. Head back up to the parking area and walk back (south) along the road.

The Holyoke Basalt crops out along the road (in places beneath poison ivy), and there have been several studies of it conducted here by geologists from Wesleyan University and the University of Connecticut. You may be able to see some small round holes left by drills that removed cores to be used in measuring the magnetic properties of the basalt. The earth's magnetism has changed over time. We can determine its direction and strength in these rocks when they cooled from liquid lava, and compare that with Early Jurassic rocks at other places.

Dark bands of coarse basalt up to two feet thick cut across poorly formed, steep columns in some of the road cuts. These bands represent magma that was squeezed out of the lava after it had partially solidified into a "crystal mush" of plagioclase grains. The mobilized liquid spread outward into these sill-like segregation sheets, and if this mechanism occurs in large enough volumes, it could explain how two different basalts can be made from one original magma.

Farther to the south (close to the power lines) the Holyoke shows better-developed cooling columns, but they are smaller, and not all straight and uniform in orientation. This type marks the upper zone within flood basalts called the entablature, as opposed to the lower colonnade zone of larger and straighter columns. The colonnade and entablature form by simultaneous cooling and solidification of the lava from the bottom and top of the flow, and in many traprock cliff faces you can see a distinct but irregular boundary between the two zones.

A little farther to the south an old road angles down to the river below a broken masonry dam. Here you can see an exposure of red siltstone of the East Berlin Formation, which is featured at some other locations because of its abundant dinosaur tracks. The East Berlin formation lies over the Holyoke and under the Hampden Basalt, and that last basalt is also exposed along the paved road closer to the highway bridge to the south. Because they are thinner than the Holyoke Basalt, the Talcott and Hampden Basalts have not developed segregation sheets.

Things to discuss: It is apparent that the lava flows are much more resistant to erosion than the sedimentary formations around them. Yet the Farmington River cut this gorge across the basalt ridges rather than just turning to flow parallel to them down the valley. How or why would the river do this? Is it possible to estimate how long the river has flowed through the gorge, based on how well developed (in steepness and width) this valley has become?

Photo: Greg McHone.

Powder Hill Dinosaur Park, Middlefield

A small brownstone quarry has been turned into a little-known Middlefield town park where you can see several large and small dinosaur footprints. The property was formerly owned by Yale University, which used it for scientific studies. In addition to the tracks, you can see good examples of East Berlin siltstone, which seems to be the formation with the most footprints.

How to get there: Take Route 66 eastward from Meriden or westward from Middletown to Route 147, and turn onto that road south toward the Powder Ridge Ski Area. In Baileyville (a tiny mill village) turn right onto Powder Hill Road, and continue about ¼ mile past the entrance road to the Ski Area to the park on your right.

What you can see: According to the experts, four different dinosaur track species have been found at this site: *Eubrontes*, *Anchisauripus*, *Grallator*, and *Anomoepus*. The first three are three-toed theropod tracks that differ only in size from large to small, which could represent either different animal species or the same species at different ages. *Anomoepus* is believed to be from a small plant-eating dinosaur. The tracks are not very distinct, but you can quickly find some of the larger ones called *Eubrontes*, the Connecticut State Fossil. It appears that fossil looters or vandals have been at work here, so we hope that the location does not deteriorate.

The layers of fine-grained sandstone or siltstone are "parted" along bedding planes, probably due to clay that washes out more easily between the layers. The parting also made it easier for the quarry workers to split out blocks of similar thickness and size, which would make them more valuable for building stones. The sediment probably was deposited in a lake, which had a bottom of fine sand of uniform size along with some mud and clay. If the water was shallow or the mud was sometimes exposed in times of low water levels, animals would leave tracks. In order for the tracks to be preserved, the mud must have become somewhat dried and hardened, but it was soon covered by more mud when the lake level rose again.

Things to discuss: Do you think that dinosaurs preferred to walk along lakeshores, or do we mainly see tracks in this type of rock because it preserved them better? The next time you are at a lake or beach, look for tracks from birds and other animals along the edge of the water. How likely is it for those tracks to become preserved in stone?

The view of Mount Higby from Beseck Mountain
Photo: Greg McHone

Beseck Mountain, Middlefield

An easy trail up Beseck Mountain has exposures of the Shuttle Meadow formation as well as Holyoke Basalt, the massive middle lava flow of the Connecticut Valley. At the top there is a fine view across the valley toward the Hanging Hills of Meriden, part of the traprock ridge system. The views and interesting ecology attract many hikers to the "Blue Trail" ridge system.

How to get there: From Route 66 in the eastern side of Meriden, exit onto East Main Street and turn east (toward Middletown). Black Pond State Fishing Access is below Beseck Mountain just past where East Main Street merges into Route 66 eastward toward Middlefield. Park in the gravel lot above the fishing access area (do not block the fishing boat ramp). Walk back past the gate and up the hill along an old road overlooking Black Pond.

What you can see: As you walk up the hill, notice the outcrops of gray shale to your left. This is the upper part of the Shuttle Meadow Formation, which represents clay-rich mud deposited in lakes that formed over the Talcott Basalt, a formation that is deeper under the ground (at about the level of Black Pond). Break up a piece of the shale: can you tell from its powdery nature that it is made from clay? Clay minerals are microscopic but very flat, and when the grains of clay settle to the bottom of a quiet lake, their flat sides create a very level layer that splits easily. This splitting characteristic is called fissility, and shale by definition is very fissile. Otherwise, it would just be claystone or mudstone. Sometimes there are fossils of tiny water animals or plants, or even forms of fish preserved in shales like this—it is worth a look.

Continue up the old road into the lower part of the Holyoke Basalt. Note the way the basalt breaks into short columns with five or six flat sides. This "columnar jointing" is caused when the lava shrinks as it solidifies and cools, cracking it into regular polygonal sides with their longest dimension perpendicular to the top and bottom of the flow. Higher up, the Blue Trail splits off the road to the right and continues to the top of the ridge. Here you can see polygonal outlines of the tops of columns, caused by harder silica minerals deposited into the joints. Off to the west and north are more traprock ridges, all made with this same Holyoke Basalt.

Things to discuss: The Holyoke fissure volcano was fed by the Buttress Dike, which during a massive eruption "flooded" the entire valley area, and beyond, with lava flows hundreds of times larger than any known in our human history. Can you imagine being close to such an eruption? What effects would its lavas, ashes, and poisonous gases have on living things? After the eruption stopped, how long might it be before plants and animals could live here again?

Photo: Greg McHone.

Talcott Basalt and Pillow Lavas, Meriden

Although shopping centers and malls are not admired by everyone, they sometimes have fine rock cuts around them with safe, quiet places to walk, unlike highways. The Mesozoic rocks around Meriden Square and the nearby Target store are much appreciated by geology students.

How to get there: From I-691 West take Exit 6, and turn left to go around the Westfield Mall and up the hill to the Target store. From I-691 East take Exit 5, and turn left to the Target store. Park by the left side of the store. You must stay on the pavement or grass along the parking areas—do not walk behind the store or climb on the rocks. This is private property and it is a special privilege to see these wonderful rock exposures. While you are here, buy something to show your appreciation!

What you can see: Most of the cut is in Talcott Basalt. Toward the bottom of the lava flow, the rock has numerous rounded structures called pillows, and beneath that is reddish siltstone of the uppermost New Haven Formation. Pillows are formed when lava flows into deep water. The water quenches the outer layer of lava as it flows, causing a balloon effect from the pressure of the liquid basalt inside. Tongues of lava swell into these pillow shapes, which sometimes break off but are soon covered by the lava with more pillows. After enough water is displaced, the flow can become more solid, as appears to be the case in the upper part of this cut.

Pillow structures characterize the base of the Talcott Basalt across much of Connecticut. The weight of the dense lava has forced the soft lake sediment up into the flow. There must have been a lake over a large area of central Connecticut 201 million years ago, because many other exposures of the base of the Talcott Basalt also show pillow structures. The silt and sand of the lake bottom has been pushed up and around some of the pillows, probably helped by steam generated by the hot lava, which was close to 2,500 °F. White minerals fill some of the gas bubbles and fractures caused by the steam.

Look up at the cliff behind the store. This is South Mountain, made from Holyoke Basalt. This is one of the "hanging hills" or traprock ridges with steep cliff faces. Faults cutting through the basalt flows have created several cliff faces like this one, in which the irregular boundary between the lower colonnade and upper entablature is visible. Basalt columns in the entablature are typically twisted and turned into different angles, unlike the more regular and uniform columns of the colonnade. This difference is thought to be caused by rainfall seeping downward into fractures in the top of the lava flow. The water in the fractures caused the columns to cool more rapidly toward those places, which deformed them. The lower colonnade did not have this problem because there are no rain-filled fractures in the lower part of the lava flow.

Things to discuss: Could this lake have been shallow, or was it deeper than the thickness of the lava flow? Would you expect to find lots of fish fossils here?

Photo: Nancy McHone.

Castle Craig, Meriden

A popular picnic spot (in warm weather) is the top of East Peak, where a stone tower called Castle Craig provides a great view of the Connecticut Valley. Here is the place to appreciate the great lava flows that covered this region.

How to get there: From I-691 take Exit 4, and turn eastward onto West Main Street toward downtown Meriden. Part way down the hill you will make a left turn into Hubbard Park. Drive around the pond, past the swimming pool, and left to go through a gate and under I-691 onto a narrow paved road. Drive about two miles past Merimere Reservoir and up East Peak (bear left where the road forks) and into the parking area at the summit below Castle Craig.

What you can see: The rocks on the summit around the stone tower have been smoothed and rounded by glaciers during the Ice Age, which melted away only about 16,000 years ago. You can tell which way the ice moved by the long axis of the smoothed ridges, in this case about north to south. The tops of basalt columns developed in this section of the Holyoke Basalt (near the top of the entablature) are outlined into interesting polygonal patterns by low ridges of resistant minerals, possibly rich in quartz, and lines of the same minerals in straight fractures.

On a clear day, you can see ridges to the south called East Rock, West Rock, and Sleeping Giant. As described earlier in this chapter these are eroded sills of diabase of the same magma as the Talcott Basalt and Higganum Dike.

To the west you can see across the western border of the Hartford Basin into Iapetos Terranes (the Connecticut Valley Synclinorium) of the western uplands. To the east, the nearby traprock ridges stand high with their slopes of broken basalt blocks called talus. The very farthest ridge line is in the Bronson Hills Terrane, across the eastern border fault. The nearby traprock ridges are all made of the same Holyoke Basalt Flow as at Castle Craig. Since the flow tilts down to the east, yet reappears in several ridges, there must be faults between the ridges that have offset the layers to lower levels as you go from east to west. The realization that faults could explain how the same strata repeat along the surface was first made in the 1880s by William Davis, a geologist from Harvard University who studied these same Hanging Hills of Meriden.

To the north, more traprock ridges show the steady eastward tilt of the strata more clearly. At least one quarry is visible, a place where machines convert the hard but brittle basalt into crushed stone. This is a critical material for the construction of buildings and highways. However, there is strong sentiment to preserve our traprock ridges, and it will be difficult for many new quarries to open after these older ones become depleted.

On your way back through Hubbard Park, stop to see the small collection of dinosaur tracks in large boulders between the road and the eastern side of the pond. You may be able to identify *Eubrontes* and perhaps other ichnospecies.

Things to discuss: There are a fair number of water reservoirs around traprock ridges in Connecticut, including the one you passed on the way up to Castle Craig. Is there any advantage to having them in such places? Think about why there are relatively few roads, houses, farms, and factories in the traprock ridges, and how less development means cleaner water supplies in these "recharge areas."

Photo: Greg McHone. Flight courtesy of Larry Pond.

Platt Farm Preserve, Southbury

In recent years several geologists have turned their attention to the Pomperaug Basin, a small area of Mesozoic rocks in Southbury and Woodbury. Although much smaller than the Hartford Basin, the Pomperaug has a similar package of coarse Triassic sandstone under Early Jurassic basalts and mudstones. If you live in southwestern Connecticut, this terrane provides some nearby and interesting Mesozoic field sites. One area that is especially easy to visit is the Platt Farm Preserve, a town-owned conservation property near the village of South Britain that is not far off the interstate highway in Southbury.

How to get there: The Platt Farm Preserve is easily reached from I-84. Take Exit 14 (South Britain exit), and head north on Route 172. As you enter the village, turn right at the old library onto Library Road, and after a few hundred feet turn right onto Flood

135

Bridge Road. In less than a mile there will be a left turn (north) onto a road that looks like a driveway between a farmhouse and a barn (see the air photo). Drive past the barn and up the hill into the gravel parking lot.

What You Can See: Hiking trails branch out in several directions from the parking lot, but some of the best outcrops are along Cass Brook, which runs below the bank to the east of the parking lot. Walk northward along the lower trail behind the wooden preserve sign. Where it turns downward into the woods to cross the brook, walk downstream to see the South Britain Arkose, a fluvial (stream-deposited) sandstone that is the same age as the uppermost Late Triassic New Haven Arkose of the Hartford Basin. The sedimentary layers dip eastward, just like sediment beds in the larger basin. Faults along the eastern margin of each basin have tilted the rocks downward, and erosion has created valleys in the softer faulted and sedimentary areas.

Continue back upstream along Cass Brook to see weathered basalt along the stream banks. This is the East Hill Basalt (a preliminary name for this newly studied formation), which is the same lava as the Talcott Basalt behind the Meriden Target store, although it is much thinner here. The fact that the basalt is less than 30 feet thick at Platt Farm is not just due to the distance from its fissure source, the Higganum Dike (described above at Hurd State Park), but also because of a higher elevation of the land in this direction. Lavas in these basins appear to have flowed 60 miles or more from their sources.

After a few bends in the stream, there is an eroded gulley high up in the right (east) bank above the brook, which you should climb up to see. The lower part of the gulley exposes gray-green to reddish shale of the Cass Brook Formation, which was deposited as mud in a lake that formed directly over the East Hill Basalt. As in the Hartford Basin, it is likely that some lakes were formed because the lavas had blocked the drainage system and "leveled the land." Higher up in the gulley there is an abrupt change to an unusual rock called reibungsbreccia, which is a mixture of basalt blocks surrounded by a red siltstone. This rock is thought to be formed when solid basalt was faulted against soft silt (not yet siltstone), which flowed around the broken pieces of basalt. The tectonic event must have occurred fairly soon after the mud was deposited. You might be able to see some of these faults as sharp straight divisions in the gulley rocks. Farther up the gulley the reibungsbreccia grades into coarse arkose.

If you go back to the parking lot and head west (away from Cass Brook), the trail wraps around Sherman Hill, a small peak of Orenaug Basalt that is comagmatic with (the same lava as) the Holyoke Basalt of the Hartford Basin. Most of the ridges of the Pomperaug Basin are made of this basalt, which here is up to 300 feet thick, or about 2/3 of the thickness of the Holyoke Basalt. Note the well-developed cooling columns, a characteristic feature for all flood basalts. Another place where you can see this basalt is in Orenaug Park, on the east side of Woodbury Village a few miles to the north.

Things to discuss: Today there are hills between the Pomperaug and Hartford Basins. How could the lavas flow here from fissure sources farther east? Could the Pomperaug Basin have once been within the southwestern part of the Hartford Basin, before it was eroded? What other evidence would help to answer that question?

The Ice Age in Connecticut 6

Photo: Nancy McHone.

A s ancient as the rocks are below ground in Connecticut, the materials and landscapes at the state's surface are relatively young. The topography of the state (the physical features of the earth's surface) was transformed by the great sheets of ice (glaciers) of the Pleistocene Ice Age, which ended just 10,000 years ago, when the last of continental ice sheets remaining in northern Canada and Europe finally melted.

Our hills were carved, rounded, and cut down under the thick ice formed during the deep freeze that began 2 million years earlier. As the ice melted locally, it filled our valleys with thick deposits of gravel, sand, and clay. This glacially sculpted landscape and leftover boulders and gravels account for much of what is familiar about the earth forms seen in Connecticut today.

Retired State Geologist Ralph Lewis tells a story about how for generations the glaciers even shaped peoples' habits. Before Interstate 95 was built to provide a relatively smooth route along the coast from Greenwich to Stonington, people traveled east or west across Connecticut less frequently. It was almost always easier to travel north or south instead, along routes formed naturally through the long valleys, than it was to struggle over the corrugated hills across the state or around the coves that line the coast.

The fact that travel in Connecticut is still mostly north to south—even with the interstates—is due largely to the work of the glaciers of the last Ice Age. Thousands of years ago, as the ice grew in the north, gravity worked to push glaciers in southerly directions. The direction of the ice corresponded with the direction in which rock structures of the metamorphic terranes were folded and wrinkled during ancient mountain-building events. As the ice pressed south, it eroded land and solid rock alike, emphasizing the north-south grain of the state even more. Older roads and state routes commonly still follow these convenient pathways between the hills, while the best views are usually found cresting hills going east-west.

The work of the glaciers is apparent on smaller scales as well. Surprisingly smoothly polished bedrock, known as ledge, can be found throughout the state. These are exposed today because glacial soils are often thin and are easily eroded from hillsides. Look at the surface at a low angle and parallel groups of long scratches and grooves are often seen. These are striations, scars cut in bedrock by stones frozen in the base of the glacier. As the ice moved, these stones scrubbed the surface like an enormous sheet of sandpaper. More than just curiosities, geologists use striations today to measure and plot the directions in which the ice moved, and have found it was almost always from the northwest, north, or northeast.

Till and stratified drift

Beginning about 16,000 years ago, the thick layers of ice that had expanded to cover New England began to melt. The ice melted back to the north until Connecticut slowly reemerged. As with all glaciers, the ice that covered the state contained a vast amount of powdered rock, silt, sand, stones and boulders of all sizes—materials glaciers scraped from hillsides and landscapes further north and carried south.

As the ice slowly withdrew, this material (known collectively as till) fell where it stood. Till was deposited over the hills in layers from a few feet to tens of feet deep. This glacially transported sediment is easy to identify by its great assortment of different materials and rock of all sizes.

Layers of till do not make good local subsoils. Farmers struggled with what they came to know as "hardpan," and the stones that seemed to grow out of fields each year following spring thaw. Many were piled in the stonewalls found everywhere in Connecticut. The poor sorting of the till materials cause them to pack tightly together, making it difficult to dig. Till is also known to conduct water poorly (known as low permeability).

Drift refers to other kinds of glacial sediment. Drift is dumped in front of and under large, melting glaciers, where there is plenty of meltwater flowing. This water can carry sand and gravel from till into lakes that often collect nearby. Depending on the energy of a stream, sand and gravel are transported at varying rates and deposited in alternating layers to form deltas that extend into these glacial lakes. At the top, a delta may start out fairly level, but can change as the land beneath it rises (rebounds) after being depressed under thick, heavy ice sheets. The material deposited within the delta is known as stratified drift, in keeping with its layered nature.

A large delta deposit of stratified sand and gravel exists today in Rocky Hill, where it has been quarried for years. This delta was part of a natural dam for a once vast glacial lake, named for its discoverer, the pioneering geologist Edward Hitchcock. Stratified drift fills many valleys, large and small, in Connecticut, including the Connecticut River Valley. Today, the hundreds of quarries operated to dig this sand and gravel form a large industry. Many towns and cities rely on major natural water sources (or aquifers) formed in what are deposits of stratified drift.

Stratified drift has also preserved important clues to environmental change that occurred during the time the ice moved out of Connecticut. Fragments of wood and bone buried in glacial sediment have since been carbon-dated to determine how long ago various plants and animals lived. This evidence has revealed that by 11,000 years ago, the barren landscape left behind by the glaciers was a tundra-like environment, covered by cold-region forests and inhabited by large animals, from horses and mastodons and deer to the humans found nearly everywhere in Connecticut today.

Paleontologist Spencer Lucas has written about discoveries of Pleistocene mammals preserved in Connecticut in a variety of glacially formed sediments. According to Lucas, Benjamin Silliman reported that teeth from what is now thought to be a mastodon (a large mammal related to mammoths and elephants) was found in 1827 near Cheshire. Mastodon bones were also found in New Britain in 1833 and in Sharon in 1835. The antlers of a deer were unearthed about 1850 in North Haven and the bones of a caribou were discovered at Quinnipiac in 1875. More mastodon evidence turned up in Bristol in 1885, and the famous Farmington Mastodon, a nearly complete skeleton, was found in 1914. A Pleistocene horse was also found along the bank of the Connecticut River near Glastonbury sometime in the 1920s or 1930s.

Moraines, eskers, kettles, and drumlins

The term "moraine" comes from the French word used to describe glacial features observed in the Alps. It refers to glacial till that builds up along the ice margin into a ridge that essentially outlines the glacier, whatever its size. Billions of tons of sediment were transported within and on the continental glaciers. Ice constantly flows to its margin (or leading edge). As it melts, sediment falls out at the margin as till or drift. If the margin itself is changing, the till may be spread into a fairly thin layer. If the rate of ice flow is matched by the rate of melting at the margin, the end of the glacier may appear relatively stationary. This is known as "stagnation" even though the ice may be continuing to flow.

Water also melts along the margins beneath a glacier, where it can cut tunnels in the ice even before reaching daylight. Sediment can melt out of the interior of the glacier to fill a tunnel and form what is known as an esker. The sediment remains as a ridge, curved in the shape of the stream that had once flowed beneath the ice. This ridge, or esker, may look like moraine, but runs at an angle to the ice margin rather than with it.

Kettles are formed from blocks of ice that break off a glacier and are covered by sediment that collects along the margins or even beneath the glacier. These blocks often do not melt until after the glacier has left the area. When they do, the land surface that accumulated over the block sinks to form a bowl-shaped depression called a kettle. Kettles can be as large as hundreds of feet across and several tens of feet deep and they may or may not hold ponds of water. Kettles may form in eskers, stratified drift, or thick ground till.

Of course, the glacier may melt back only to advance again, so that it rides up over older deposits of glacial sediment. These piles may become streamlined into tear-drop shaped hills and ridges called drumlins, in which the up-flow end is blunt and steep while the down-flow end is stretched into an elongated tail. There are many drumlins in Connecticut. Among the best known drumlins are Horse Barn Hill at the University of Connecticut campus in Storrs and Indian Hill in Middletown.

Glacial Lake Hitchcock

About 18,000 years ago, the southern end of the last ice sheet retreated northward out of Connecticut. This does not mean that ice stopped flowing from Canada into New England, but rather that the glacial margins melted faster than the ice could move southward.

During this retreat, glacial Lake Middletown formed in the Connecticut River Valley (at Middletown), which left its own clay and sand deposits. At Rocky Hill, a large, deep delta of sand and gravel was deposited where glacial meltwater flowed into Lake Middletown. When the ice margin retreated farther

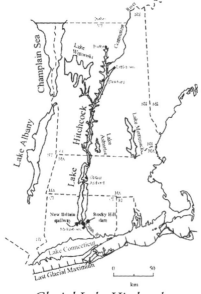

Glacial Lake Hitchcock
Courtesy of Dr. Julie Brigham-Grette, Dept. of Geosciences, University of Massachusetts, Amherst

north and Lake Middletown became lower, this thick sediment delta became a wide dam in the Connecticut River Valley. Water trapped between the continental ice sheet and the Rocky Hill dam created glacial Lake Hitchcock about 15,000 years ago. As the ice continued to melt northward, this lake expanded up the Connecticut River Valley. At its maximum extent, glacial Lake Hitchcock extended for 200 miles up the valley from Rocky Hill to St. Johnsbury, Vermont, and also up valley branches on either side.

The water level of Lake Hitchcock changed several times. Initially, Lake Middletown and Lake Hitchcock were connected, and their water level was controlled by the Lake Middletown spillway, which drained the lake to the southwest toward New Haven. A subsequent drop in water levels separated the two basins, and a new exit for Lake Hitchcock formed in a low divide near New Britain, called the New Britain Spillway. The new water level slowly lowered as glacial till in the spillway channel was eroded. A more stable phase of Glacial Lake Hitchcock began after the New Britain Spillway reached the bedrock level, which prevented further erosion and lake level change. The lake remained at this level about 2,000 to 3,000 years while the ice retreated from Connecticut into northern Vermont. Around 12,000 years ago the Rocky Hill dam broke, which must have caused quite a flood as glacial Lake Hitchcock drained down the ancient pathway along which the Connecticut River flows today. Glacial Lake Hitchcock was long but narrow where it filled the Connecticut River Valley. The map of Lake Hitchcock was developed by Dr. Julie Brigham-Grette of the University of Massachusetts, a researcher and authority on glacial lakes.

A portion of the sand and gravel delta deposited at Rocky Hill
Photo: Greg McHone

Glacial Lake Connecticut

During the time that New England was covered in ice, the level of the Atlantic Ocean was about 200 feet lower than it is today. When the ice sheet first melted back from its moraines along the south shore of Long Island and into Connecticut, about 18,000 years ago, a large fresh water lake called glacial Lake Connecticut filled the area that today is Long Island Sound. The lake water was dammed by high ridges of moraines that still exist today in eastern islands such as Fishers Island.

Lake Connecticut was deeper and colder than today's salt-water sound. When it was new, icebergs likely floated on its waters, and its color was cloudy-white from "rock flour" ground from bedrock by the glacier. About 3,000 years after it originated, Lake Connecticut drained through an eroded gap in the moraine ridge near Fisher's Island. The lake bottom became dry land with trees and animals for about 1,000 years, but as the great ice sheets continued to melt, the Atlantic Ocean rose and gradually filled the sound. The sound is still rising today at a rate that seems to be increasing, possibly because global warming is melting some continental glaciers in Greenland and Antarctica.

The rising sea level will soon become a serious problem for many cities that are not far above sea level, as well as for beachside communities in Connecticut.

Photo: Greg McHone.

Hammonasset Beach State Park

Hammonasset Beach State Park is extremely popular in summer because of its two miles of beach along the north shore of Long Island Sound and large campground with over 500 sites. But even on a busy summer day, the eastern end of the park is a good place to see how glaciers have shaped the coastline.

How to get there: Hammonasset Beach is reached via its own four lane road from Exit 62 of I-95 in Madison. Off-season the park is free, but during the summer there is an entrance fee. Drive in and around the large salt marsh on your left, staying left at the traffic circle, and heading east to the end of the road at Meigs Point.

What you can see: To the north of the road there is a large salt marsh containing a variety of plant and animal life. At Meigs Point, a long ridge of large boulders separates the marsh from the water. This is the Hammonasset Glacial Moraine, left by the melting ice as a long ridge of sand, gravel, and rocks, which was then sorted by waves. A smaller moraine to the north, separated by marsh, is covered with trees.

The Meigs Point trail follows the top of the moraine. A trail which used to go to Cedar Island goes through a more open forest growing on a smaller moraine to the north of the marsh, ending at an observation platform in the marsh. Because of the various habitats along these trails, a wide variety of plants and animals will be seen. This marsh area is especially good for viewing migratory and wetland birds. Much of the following description is provided by Nancy McHone of the State Geological and Natural History Survey.

Human history
When European settlers arrived in the mid- to late-1600s, the area around today's park was being farmed by the Hammonasset Indians, whose name means "where we dig the ground." Their main crops were probably beans, corn and squash, plus they also collected shellfish and fished in the Sound and rivers. The Hammonasset Indians later turned over the area around Hammonasset to the Mohegan tribe as part of a marriage dowry. The Mohegan Sachem Uncas sold the park area to George Fenwick in 1639. Fenwick later gave or traded this land to the Guilford colony for use as farmland. The colonists mainly used the area to gather seaweed and to cut salt-marsh hay for feed and bedding for horses and cattle.

The eastern end of the current park was once owned by the Winchester Repeating Arms Company, which had been using the property as a testing range since 1898, especially before and during World War I. The state asked the arms company to give first right of purchase of the land to the state. When Winchester decided to sell the land in 1923, after being approached by a hotel and resort developer, they contacted the state. Unfortunately, the state did not have the money available at the time. Luckily, a member of the Park and Forests Commission, Mr. J. Harris Wittemore, stepped forward and purchased the land with his own money, leasing it to the state for five percent of the purchase price plus taxes. Two years later the state bought the land for $63,000, and the park officially opened on July 18, 1920 as Connecticut's 19th state park.

During World War II, the park was leased to the War Department and closed to the public. Warplanes used targets set up along Meigs Point Road for target practice, being careful to shoot toward the Sound rather than toward the shore. One plane crashed during practice and remains in the Sound today. In 1963, a man using a metal detector to search the beach for coins turned up a World War I mortar round. The park was closed for a time while the beach was searched for more old munitions, but none were found.

After the war, the park returned to public use, although some repairs were needed first. The War Department provided some repair funds, the state coming up with the rest. After the 1955 hurricane, a stone jetty was built at Meigs Point to help protect the beach from erosion. However, because sand is carried along the shore from west to east, the beach is becoming narrower on the east side of the jetty because much of the sand is trapped by the jetty.

Geological history

Around 21,000 years ago, Connecticut and Long Island Sound were completely covered with a layer of ice at least a mile thick. Glaciers pick up a variety of silt- to boulder-sized materials as they move over the ground. These materials were left behind as the glacier melted, creating Hammonasset Beach State Park. Winds and waves have continued to sculpt the area since that time. The Wisconsin Glacier first advanced into Connecticut from the north about 26,000 years ago. The ice front reached as far south as the southern edge of Long Island in this area, before its advance stopped. Then as the summer melt exceeded the winter snowfall, the glacial front gradually receded to the north, even as the ice continued to move south. The melting ice left behind large piles of rock debris carried by the ice. Sometimes the ice front was stationary for a longer period, causing larger deposits of glacial material to build up. These long, linear piles of unsorted clay- to boulder-sized materials are called moraines.

A little less than 17,500 years ago, the ice front paused on a line from Hammonasset through Ledyard to Queens River, Rhode Island, depositing the Hammonasset-Ledyard-Queens River Moraine, a double moraine. This is the moraine so well displayed at Meigs Point. The southern moraine starts to the east and continues northeast along the shore as a string of uneven, low rises,

Photo: Greg McHone.

with various breaks, until disappearing into the entrance to Clinton Harbor. A trail follows the ridge top of the moraine. Along the water's edge all that remains of the moraine are huge boulders where wave action has removed all but the largest rocks, the ultimate coarse-grained beach!

As you follow the trail, look at the variety of materials in the ridge, ranging in size from microscopic clay particles to huge boulders. Moraines typically are made up of mixed materials of various sizes. From below the ridge top, down among the boulders, you can look back at the ridge in some places and see the interior of the moraine exposed. There are also smaller rocks as well as gravel, sand and silt-sized particles. This is how the materials were originally dumped by the melting ice, all in a mixed jumble.

The Meigs Point Trail runs along the top of the south moraine, over large rocks, then down to a boulder beach where there is a break in the moraine.

The southern moraine is especially interesting because the waves keep eroding the Sound side, allowing the internal structure of the moraine to be exposed. Where waves have removed the smaller materials, boulder lag beaches remain, with areas of boulders scattered along the water's edge. The longshore currents have carried the finer materials from west to east, forming the beach out to Cedar Island.

Things to discuss: Is there any reason why the glacier should have stopped here for a while, or is it just a coincidence that the moraine follows part of the Long Island Sound shoreline? Why is the beach mainly just sand while the moraine next to it contains so many larger rocks?

Photo: Greg McHone.

Glacial polish and striations, Essex

You can find glacially smoothed and polished bedrock surfaces in most of the state, often with parallel grooves or striations made by the moving ice. Near Essex is an excellent example.

How to find it: From Route 9 in southern Connecticut, take Exit 3 and turn south onto Route 154. Drive south a few tenths of a mile and turn left into the driveway of the office park. Park near the eastern end of the driveway and walk along the building to see the rock face exposed on the hillside. A glacially smoothed surface with striations is very well displayed on this hillside in Essex.

What you will see: This outcrop of gneiss in the Avalonian Terrane has had its thin soil eroded, leaving the bare rock surface that has been smoothed and sculpted by glacial ice. As the ice slid over the surface, it also plucked rock out of some sections, leaving rough irregular surfaces in those places. Surfaces of hard rocks such as these can become very polished, so that when wet they look shiny.

Even harder rock materials must make up the stones that were pressed against the surface under the ice as it moved, creating the grooves and scratches that run along the side of the hill.

Things to discuss: Can you estimate which way the ice moved at this location? Are these just lines with two ends, or can you tell the one direction that ice moved along the lines?

A view of Salmon Cove from Mount Tom
Photo: Greg McHone

Echo Farm glacial drift, East Haddam

Echo Farm was purchased by the state in 1999 as open space. The land may eventually become a state park, but it is essentially a nature preserve. The former owner constructed a network of gravel roads leading to scenic overlooks, ponds, meadows, and oak forests with much of the understory (brush and smaller trees) cleared away. Despite its name, the preserve has not been farmed for years. Wildlife is abundant at Echo Farm and visitors have reported seeing turkeys, hawks, eagles, many songbirds, deer, bobcats, and other small animals.

Geological features include drift (till) and stratified drift (sand and gravel) deposited by glaciers, bedrock ledges of Hebron

Gneiss, and pegmatite. The sand deposits and pegmatite have been quarried during the last century, so they are well exposed and accessible. Landforms of bedrock hills and river valleys are very well displayed at the scenic vistas.

Legend has it that Mount Tom (the highest hill at Echo Farm) and Cave Hill to the northeast were used by Native Americans for special gatherings and spiritual ceremonies. The area appears to be a center for the mysterious "Moodus noises," which sound like distant thunder or cannon fire when they occur (very rarely in recent years). Although we now know that very small earthquakes cause the sounds, in former times some superstitious people blamed "Indian witches" or evil spirits. The last Native Americans to leave the area supposedly warned the European settlers to treat Mount Tom with special respect. Thus, we might feel that it is very appropriate to preserve Echo Farm in its natural state.

How to get there: Head north from Goodspeed Landing in East Haddam or south from Cobalt in East Hampton along Route 155. Echo Farm is south of Sunrise Resort in the northern part of East Haddam. Turn west onto Echo Farm Road from Route 155, park near the gate at the end of this short road, and walk in.

This is a nature preserve, not a public park, so let other people know where you are. Before large groups visit, their leader should contact the Eastern District Headquarters for state parks.

Walk past the ponds and turn left to go up the hill and into the oak woods of Mount Tom. As on all state properties, collecting or removing any rocks, minerals, plants or other materials is not allowed without a permit. The best idea is to not collect anything but photographs. To help with your tour, bring along field guide books to identify birds, trees, insects, flowers, and minerals that are so abundant and interesting here. The view to the southwest from Mount Tom shows Salmon Cove and the hills across the Connecticut River, all of which were modified by glacial ice.

What you will see: There are two vistas on Mount Tom. The first is along a side trail to the left after you reach the top of the hill. Here you can look out across Johnsonville, a collection of buildings around a mill pond where a twine mill once operated. The pond is on the Moodus River, which provided water power in the nineteenth and twentieth centuries for a series of mills that employed

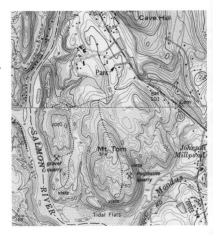

many people in the area. The pond is held by one of the last large timber dams left in the state. Valleys such as the one the Moodus River flows through were deepened and widened by the movement of the glaciers. The view up the Moodus River Valley is into the upland hills of Connecticut, which were rounded and lowered by the ice as it flowed southward. As the ice melted, the valleys were filled in some places by ice-contact sediment, and eroded in other places by rushing meltwater streams. The second vista is at the end of the trail to the south. Here you look across Salmon Cove, a tidal basin where the Salmon River meets the Connecticut River. The low hills around the cove and across the river are likewise sculpted by the ice.

Till and drift

The glacial till that mantles the ground is called drift, which is very evident in fields along the gravel road where the top soil was scraped off by the previous landowner. Most of the rounded cobbles and stones in the till are made of gneiss, the most common rock type in Connecticut. Gneiss is a hard metamorphic rock with bands of lighter and darker minerals. Note that the road itself was partly surfaced with gravel that is not from the local area, including angular pieces of gray traprock (basalt), an igneous volcanic rock from the central Connecticut Valley farther west. Although Connecticut gneiss is mostly 300 million to 600 million years old (Late Proterozoic to Paleozoic), the traprock was all formed in volcanic eruptions about 200 million years ago, or in the Early Jurassic Period (Early Mesozoic).

Stratified drift

If you go back down Mount Tom and follow the gravel road where it branches off to the west, you will eventually come to some small quarries in sand and gravel that were used by the former owner. Some of these are obviously made of till similar to what you have already seen, but in addition, there are layers mainly made of sand as well. The flow of water stratified, or separated, glacial till into distinct layers of sand and gravel, and larger boulders were left behind while finer silt and clay were carried away. As the glaciers retreated, small lakes formed in many of the valleys of Connecticut, and the action of streams entering the lake water sorted out the different sizes of sand and gravel. Just beyond the quarry is the Salmon River, which flows through and over a lot of stratified drift. Many small sand and gravel quarries in Connecticut are used to provide fill for various local construction projects.

The edge of a gravel quarry at Echo Farm reveals the thinness of its top soil, which most plants and trees must have to grow well. Early farming practices led to erosion and thinning of the top soil, often revealing the stony till or drift underneath. This is why many farmers eventually stopped trying to plow their land and converted what they could into sheep and cattle pastures.

Photo: Nancy McHone.

Ledyard Glacial Park

The Hammonasset Moraine continues toward the east-north-east into the Ledyard area of southeastern Connecticut. However, it contains quite different-looking boulders at Ledyard Glacial Park, which you can see in an interesting walk through the woods.

How to get there: From Exit 88 off I-95 in Groton, take Route 117 through Ledyard Center, turn left onto Route 214, and left onto Whalehead Road. After the road curves around to the right, park along the side of the road near the power line and take the Glacial Park trail.

What you will see: A fantastic collection of room-sized boulders fills the woods along the moraine trail. Unlike the pink granitic gneiss moraine at Meigs Point which is mixed with other rock types in all sizes down to sand and silt, here there seems to be only large boulders of gray gneiss.

Most of the boulders are the same rock type as the local formation of Avalonian Gneiss, and so they could not have been transported very far. However, they are rounded enough to indicate erosion by ice and/or water, as expected. It is believed that glacial meltwater removed the finer materials that are abundant in other moraines. Bedrock of similar gneiss underlies the boulder piles as well, but there are a few examples of different rock types mixed into the boulder piles, as expected for ice-contact sediments. Immediately to the west the boulders of the moraine are dispersed.

Things to discuss: This area may be pretty obvious, but do you think it might be tricky to map moraines? Is this a feature that could be covered over by trees and soils? How is it different from an esker?

Bristol kettles

Birge Pond Recreation Area, also called "The Hoppers," was purchased by the city of Bristol from Bristol Savings Bank in 1973. It is forested open space that includes several large and small kettles (locally called hoppers), esker ridges of glacial gravel, and hiking, walking and bicycle trails. Fishing and canoeing are allowed on the pond. Quarries have removed large volumes of sand and gravel in this area, destroying some kettles in the process. This was a place where some of the continental ice sheet was stranded, so that large blocks of ice became buried by sediment in ice-dammed lakes along the edge of the glacier.

How to get there: Birge Pond is accessed from Beech Street, on the north side of the City of Bristol. Turn off Route 69 near Jennings School and head west toward the pond (see map). From the parking area, walk west along the trails and up onto the esker ridge. The largest kettles (indicated by the black-lined box on the map) are to the northwest, and if you do not wish to walk so far, try parking near the area labeled as "gravel pits." Some of the kettles and the developed areas are on private land, which you should avoid.

What you can see: The kettles are circular or oval enclosed depressions, some of which are large enough to

A topographical map of the Birge Pond area. The location of the largest kettles is indicated by the black-lined box in the upper left.

contain trees. If you find one of the old gravel pits, you can see that the sand and gravel deposits are very thick here, possibly because of a natural topographic trap from the nearby western border of the Mesozoic Hartford Basin.

If you can, scramble down into one of the kettles. Here you can better picture the large mass of ice buried by the sand and gravel. Exactly the same event occurs at the ends of large modern glaciers. The narrow esker ridges that connect some of the kettles were formed in stream tunnels beneath the ice sheet, which filled up with sediment melting out of the ice all around the tunnel.

Things to discuss: Can you picture what this area looked like when the kettles were forming? It was much colder then and there were no trees, only tundra and subarctic plants. The trees that later grew here preserved the eskers and kettles from erosion, which could still happen if they were all cut down. Can you see any recent erosion along the trails?

Coastal Features 7

Photo: Nancy McHone

Along most shorelines, sand is in constant motion. The beach you sit or play on one summer is not the same beach that was there the year before. Instead, coastal sand is carried to and from beaches continually, by waves and water currents and by coastal rivers and streams. Weather from wind to rainstorms and hurricanes moves sand around as well.

Fortunately, around tidal waters, what goes out also comes back in—and beach sands are no exceptions. Most local beaches benefit from an annual cycling of sand; what is removed under the harsh conditions and high seas of winter storms is often replaced by new sand carried in and deposited on beaches during summer lulls.

This cycle has been complicated in recent years by the encroachment of civilization along the Connecticut shore. As nature tries to move sand slowly (most of the time) or rapidly (during storms), man-made features or obstacles often interfere. Roads, houses, docks, jetties or other buildings are often very close to the water, and may even be built directly on former marshes and beaches that have been filled and graded.

Natural catastrophes also seriously impact beaches. One such event was the hurricane of 1938. The storm seems like it was long ago, but the devastation it caused keeps it fresh in the minds of all who lived through it. The storm wreaked havoc along our coastal areas before moving inland to cause tremendous damage throughout western New England. Thousands of buildings along the coast and near tidal rivers were damaged or destroyed and hundreds of people were killed. Places such as Napatree Point, just over the state line at Watch Hill, Rhode Island, were forever changed. Once filled with summer cottages, the storm took many out to sea. None were reconstructed and Napatree Point has since been preserved as a natural area and left to return to the way it was before the hurricane.

Portions of the following descriptions were developed based on teachers' workshops written by Kaye Sullivan and Nancy McHone. Historical information was provided by the Connecticut Department of Environmental Protection. Habitat descriptions were written by Dr. Robert Craig.

Shoreline features

Most people are familiar with the fine wide and sandy beaches of our state and town parks, but our shoreline also includes large sections that are mainly rocks or marshes. Although less used by people, these areas are especially important as habitats for birds and marine animals. No single section of the Connecticut coast has every type of shoreline feature, but each will have at least a few examples of classic water-shaped landforms.

Some common shoreline features of Connecticut beaches and tidal marshes include:

Barrier island: A long island that is made of sand that is parallel to the shoreline. Barrier islands produce protected waters or lagoons on their inland side.

Bay barrier: A sand bar that completely blocks the mouth of a bay. A bay barrier begins as a spit that keeps lengthening because of weak currents; eventually it may connect to the other end of the bay.

Beach: Wave-washed sediment along a coast that extends through the surf zone. Beaches can consist of shells, sand, or gravel. Winter beaches may be steeper and tend to be made of coarser particles. Beaches can change with the seasons because more erosion occurs in the winter when winds are stronger.

Dune: A hill or ridge of sand deposited and moved by wind.

Headland: A rocky point, often with a steep cliff, jutting out into a large water body.

Lagoon: A place of quiet water and fine sediment between the shoreline and a barrier island.

Pocket beach: A small crescent-shaped beach between two headlands.

Sand spit: A long ridge of sand connected to the mainland at one end, with the other end in open water.

Tombolo: A long ridge of sand connected to the mainland, like a spit, but which extends to an island at its other end.

As an example, the beach that stretches to the west of Bluff Point and alongside the Poquonock River, known as Bushy Point Beach, exists most often as a spit. During the times it does bridge to the mainland, the beach forms a tombolo.

Waves, longshore currents, and sand movements

Wind is a powerful force on the ocean, which through the centuries has been utilized to move sailing ships and may soon become a method of generating electricity. Wind also moves waves, which in the deep ocean may only appear as wide swells passing under a ship. When a wave approaches the shoreline its shape changes as the water becomes shallower. The front of the wave slows down before the back gets the message, causing the "pile up" and breaking action that we call surf. The surf zone, not far off the beach, is where the water has much more energy, causing it to pick up and carry (suspend) larger sand grains than in quieter water farther offshore.

The wind nearly always comes toward the beach at an angle to the shoreline. This angle is relatively constant because the wind

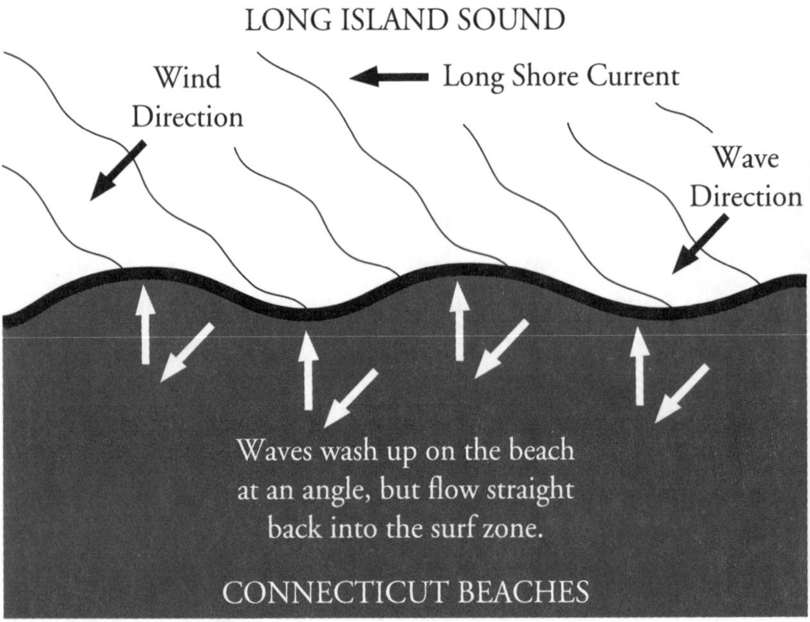

LONG ISLAND SOUND

Wind
Direction

Long Shore Current

Wave
Direction

Waves wash up on the beach
at an angle, but flow straight
back into the surf zone.

CONNECTICUT BEACHES

Schematic diagram by Greg McHone

tends to be from the same general direction most of the time (not always, of course). Thus, the waves approach the beach at an angle as well. The waves will wash up on the beach in their angular direction, but gravity causes the water to flow back straight down the slope of the beach, not back the way it came. This causes a "zig-zag" motion of the water, which also

Bushy Point Beach shows the effects of wave action in the Sound
Photo: Nancy McHone

picks up and drops sand as it moves toward and away from the beach. A steady current is produced by the waves in one direction along the shoreline not far off the beach, called a "long-shore current." The surf zone with zig-zagging wave action moves sand along the beach with this current.

During storms and high tides, most of the sand on a beach may be moved. As it is carried away, more sand usually comes along to replace it, but if the waves are strong, such replacement sand may be coarser or heavier grained, perhaps gravel-sized or even larger. This is why the stormy winter can cause a beach to become gravelly, but when the gentle winds of summer return, the finer sand can stay on the beach.

In addition, obstacles such as stone jetties may get in the way of the long-shore current. In that case, sand may move up to but not past the jetty, causing the beach to become wider on that side (perhaps good for the beachfront landowner). At the same time, sand carried away from the beach on the other side is not replaced, causing erosion and a narrower beach (and an unhappy landowner on this side). You may be able to observe this effect at the beaches described for visits.

Sea level rise

Warming of the global climate results in sea level rise for a variety of reasons. The global climate is warming due to the "greenhouse effect." Our sun's radiation is a mix of visible light, and short-wavelength ultraviolet light. Most of this radiation passes through the atmosphere except where light is blocked by clouds, and where ultraviolet radiation is blocked by ozone (a good thing). The visible radiation warms the earth's surface, which then radiates the heat as longer-wavelength infrared radiation. In the air, water vapor, carbon dioxide (CO_2), methane, and a few other gases absorb most of this infrared radiation, resulting in a warming of the earth's atmosphere. Glass in a greenhouse does essentially the same thing, so this phenomenon is known as the greenhouse effect. Without the greenhouse effect, the earth would be at least 60 degrees cooler, causing a permanent worldwide ice age.

Volcanic eruptions can release huge amounts of CO_2 as does rotting and burning vegetation, but people have added a lot more by burning fossil fuels (coal, oil, and gas) for heat and energy. The additional CO_2 and other gases released into the atmosphere are causing rapid warming of the global climate. Nature can remove extra CO_2 with new plant growth, but that takes a long time and requires more places for the plants (especially trees) to grow.

This warming of the climate causes several changes that affect sea levels. Most of the rise is because the Antarctic and Greenland ice sheets are melting, adding their water to the oceans. At the same time, the ocean water is getting warmer, which results in a slight expansion of the water, raising sea level. Also, mountain glaciers are melting.

Scientists do not agree on all of the effects that global warming will cause. Some believe that warmer temperatures will result in more water evaporation, resulting in more cloud cover, which will help to lower temperatures. Others think that the increase in evaporation will not greatly increase the cloud cover, but the water vapor will increase the greenhouse effect, raising temperatures even more. Also, the warmer ocean might release greenhouse gases that it used to keep in solution or frozen in the ocean bottom. Higher temperatures should result in less snow and ice cover, which reflects sunlight back into space. Without the snow and ice cover, more light will be absorbed, resulting in more warming of the earth. The following table lists a range of estimates for sea level rise for the next 100 years:

Worldwide Sea Level Rise Predictions, 2000-2100 (in feet)
Summary table by Nancy McHone

Scenario	2025	2050	2075	2100
Conservative	0.27	0.62	1.04	1.64
Mid-range low	0.57	1.41	2.71	4.41
Mid-range high	0.87	2.17	4.07	6.67
High	1.24	3.24	6.44	10.74

Salt marshes

The salt marshes found along the coast of Connecticut today developed following the last Ice Age, as water once locked in glaciers melted, and sea levels rose relatively rapidly. Shorelines slowly advanced further and further inland and in the process "drowned" the mouths of rivers and streams that drained the land.

Salt marshes formed in these areas by a combination of tidal action, plant colonization and organic sedimentation. Plant matter and silt gradually filled areas of the drowned rivers to create the rich, marine environments of the Connecticut coast. In places where river water mixed with seawater, estuarine tidal marshes formed.

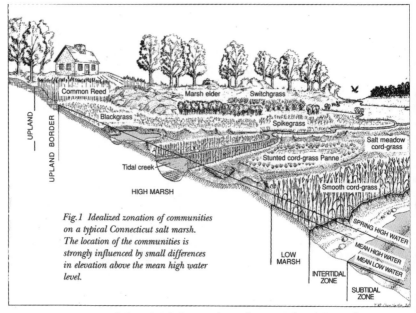

Fig.1 Idealized zonation of communities on a typical Connecticut salt marsh. The location of the communities is strongly influenced by small differences in elevation above the mean high water level.

Reprinted from Tidal Marshes of Long Island Sound
Courtesy, The Connecticut College Arboretum

Salt marshes are among the state's most important natural systems. The following description is adapted from *Discover the Critical Habitats of Connecticut,* by Connecticut biologist Dr. Robert Craig:

Salt marshes are truly critical ecological systems within the modern coastal landscape. They are notable for their high productivity (the rate of manufacture of living material), which is among the highest of temperate systems. This productivity is a consequence in large part of tidal action, which brings nutrients and minerals to the marsh, aerates the marsh's surface, and flushes away wastes.

Most of the production of the marsh is not actually consumed directly, but enters the estuarine system as detritus, or dead plant remains. Plant detritus is one of the key components of fuel for the estuary, and provides energy to power microorganism and ultimately animal productivity. The creeks of tidal wetlands are well known, for example, as a key nursery ground for a number of important game fish.

The ecological characteristics of tidal wetlands also have been associated with an ability to cleanse water of excess nutrients supplied by such human activities as sewage production. The wetlands' capacity to produce make them capable of converting excess nutrients into biological productivity. Moreover, the low energy environments of these wetlands act to filter fine sediments from water, thereby improving water clarity.

As the concentration of salt in the water diminishes away from the river mouth, a succession of plant associations appears that ranges from cordgrass at the island's seaward end to cattail and reed dominated communities at its upriver end. This beach-like area is vegetated by Dune Grass (*Ammophila breviligulata*), Seaside Goldenrod (*Solidago sempervirens*) and High Tide Bush (*Iva frutescens*).

Salt marsh is an environment characterized by plant and animal species that tolerate tidal inundation with saline water. Stands of Salt Marsh Cordgrass (*Spartina alterniflora*; in the grass family, *Gramineae*) occur in that portion of the marsh referred to as low marsh: that region from approximately mid-tide to mean high tide. It is thus found along tidal creeks and at lower microelevations. Particularly further upriver where water salinity begins to diminish, other species of cordgrass also appear, including *S. pectinata* and, less commonly, *S. cynosuroides*. These two handsome species are best located in mid to late summer when their seed heads are fully mature. One last species commonly found in this zone and occasionally predominating in it is Salt Marsh Bulrush (*Scirpus robustus*; in the sedge family).

Above this level in less frequently flooded locations is that region called high marsh. Salt Meadow Cordgrass (*S. patens*) dominates this region, and co-occurs with species like Salt Grass (*Distichlis spicata*). Other notable salt marsh specialists include Seaside Arrow-grass (*Triglochin maritima*), Salt Marsh Aster (*Aster tenuifolius*), Seaside Gerardia (*Gerardia maritima*) and Sea Lavender (*Limonium nashii*).

Toward the upland edge of many tidal marshes, Salt Meadow Cordgrass gives way to Black Grass (*Juncus gerardi*; in the rush family, *Juncaceae*). When contrasted with the pale green tussocks of Salt Meadow Cordgrass, Black Grass indeed appears black. At still higher microelevations, Switchgrass (*Panicum virgatum*), Rose Mallow (*Hibiscus palustris*) and several shrubby species like High Tide Bush come to predominate. At tidal marshes that merge into uplands, species like these form the boundary vegetation between marsh and upland.

Other notable environments in the salt marsh are salt pannes. At high tide, these surface depressions can become filled with water that does not easily drain away. As the sun evaporates their water, pannes become increasingly saline. In this "hypersaline" environment, only a handful of specialized species occur. A stunted form of Salt Marsh Cordgrass occurs, as does the fleshy-stemmed Saltwort (*Salicornia* spp.). The aquatic plant Widgeon Grass (*Ruppia maritima*; in the pondweed family, *Najadaceae*) may occur as well.

Salt marsh is often replaced by plant communities of brackish (slightly saline) water further inland. Switchgrass becomes fairly common at some spots, although Narrow-leaved Cattail (*Typha angustifolia*) and Reed (*Phragmites communis*) are the most widespread species.

Patches of salt marsh in places such as Great Island, Old Lyme, have become less widespread, apparently at least in part because of lunar cycles that influence the degree to which salt water intrudes upriver. In addition, places that were primarily Narrow-leaved Cattail in 1974 are now predominantly vegetated by Reed.

Although Reed is a native Connecticut species, an aggressive alien form has become established that replaces cattails in brackish portions of large rivers.

Rates of sea level rise have accelerated tremendously during the twentieth century and now pose a serious threat to tidal marshes. Sea levels could rise as much as four to six feet in this century. At that rate, tidal marshes will not be able to keep up. As the ocean rises, many may be drowned. Worse yet, there will be few places left where new marshes might form further inland.

Bluff Point State Park, Groton

Bluff Point Coastal Reserve offers a mix of woodland hiking and biking trails coupled with excellent wildlife viewing on Long Island Sound. The shoreline has both sandy beaches and glacially-smoothed bedrock exposures for visitors to admire and study.

How to get there: From I-91 south, exit onto Route 9 and follow it south to I-95 east. In Groton get off I-95 at Exit 88. Turn right and go to the end. Take a right turn onto Route 1. Take a left at the light onto Depot Road and go to the end. There is limited parking but no parking fees at Bluff Point State Park.

Take the hiking trail southward about a mile to Mumford Point and Bluff Point, where you will see some excellent shoreline features. Because the park has so many deer, deer ticks are abundant. Take precautions such as using insect repellent and pulling your socks up over the bottoms of your pants to keep them out. Be sure to inspect yourself after you get home.

Bluff Point is the last remaining parcel of undeveloped land along the Connecticut coast. Jutting out into waters of Long Island Sound this wooded peninsula encompasses over 800 acres. Because it is undeveloped (without rest rooms, be forewarned), and it requires a long walk in, the park is rarely crowded, although parking is limited.

History

Originally proposed for acquisition as early as 1914, it was not until 1963 that the western third of the land was purchased by the state. The park includes a north-south strip of mainland, a portion of the headland bluff fronting the Sound, and the tombolo or sand spit beach of nearly a mile in length. The beach ends at a small island called Bushy Point.

Bluff Point was designated a "Coastal Reserve" by a special act of the Connecticut legislature in 1975 to establish the area "for

Bedrock Geology of Bluff Point
Map by Nancy McHone

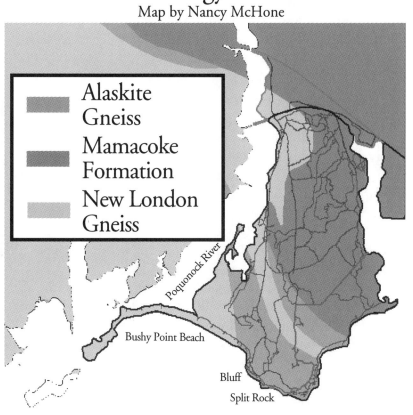

Alaskite Gneiss

Mamacoke Formation

New London Gneiss

Poquonock River

Bushy Point Beach

Bluff

Split Rock

the purpose of preserving its native ecological associations, unique faunal and floral characteristics, geological features and scenic qualities in a condition of undisturbed integrity."

Access to the bluff is by foot or non-motorized vehicle only. The trail to the bluff passes through wooded and open areas until the view broadens as the bluff is approached. Here vegetation is more sparse and diminutive because of wind exposure. Among the plants to be found at the headland are native beach plum, beach pea and red and white shore roses.

Numerous cottages once lined the shoreline at Bluff Point. All of these cottages were destroyed by the Great Hurricane of 1938. The Governor Winthrop Plantation including a farmhouse and outbuildings was located here in the early-1700s. The once-large, open fields are now grown into forests.

Geological features

As you make the walk from the parking lot and through the forest you will see boulders here and there. These were left behind by the glaciers which once covered all of New England. As they slowly moved south, loose soil, sand, gravel, and boulders were frozen into the ice. When the ice eventually melted, this material, known as moraine, was left behind. In valleys, water moved the sediments around, sorting them by the action of the moving water. As a result, sand and gravel deposits are found in valleys and New England till, a mixture of materials just sitting where it dropped, covers the hills.

Near the point the trail begins to open up and the beach appears to the west. The long walk (only at low tide) on the sandy spit out to Bushy Point Island is on sand that has been carried here by long-shore currents. The sandspit beach would probably continue to the Groton airport area except that the Poquonock River moves the sand out into the Sound. At one time the beach did extend to Bushy Point, but the 1938 hurricane breached the dunes, separating the beach from what is now Bushy Point Island, at least at high tide. The beach is now a spit, but when it was connected to the island it was a tombolo.

The Poquonock River is a "drowned" valley. This valley once led down to the wide, dry area that became Long Island Sound. After the glaciers of the last Ice Age melted, the sea rose to flood this valley along with others like it along the coast.

Sand at Bluff Point tends to move from east to west. Waves and wind work to sort sand and shells by size. Keep an eye out for areas of red or black sand. The red sand is garnet. The black sand is magnetite, which gets its name from the fact it is attracted to magnets.

Photo: Robert Craig

East of the beach and around the Point are the cliffs and large exposures of the hard, metamorphic rocks of the Avalonian Terrane. The bluffs at the southern end of the park have surfaces that were smoothed by the movement of the glacier. Where the soil has been eroded away, this surface has been exposed. It looks about as fresh today as it did 14,000 years ago.

This gray and pink-banded rock is the Hope Valley Alaskite Gneiss. It is 600 to 700 million years old and contains feldspar, quartz and plagiocase. All the bedrock here is of similar age, but varies in chemical composition.

Split Rock is the large, split boulder sitting on top of the nearly flat surface. It was probably moved to its present location by the glaciers, but because it is the same rock type as the rock it sits on, it is not considered to be a glacial erratic. On the other hand, the large boulder called Sunset Rock (follow the sign to see it) is probably an erratic, as it seems to be a different rock type than should be in this area, although the lichens make it difficult to be

Photo: Nancy McHone

sure. The rock got its name from the sunset views that used to be visible from here when this area was covered with cottages and open fields.

Surrounding Split Rock is a large exposure of metamorphic rock. Evidence that these rocks were formed under great pressure and high temperature deep beneath the surface is easily seen. Folding is apparent from the swirls of black and white layers. Veins of one or more minerals such as pink feldspar and white quartz can be seen cutting across other layers. At least one old fault is also exposed here; the offset visible in the colored layers reveals where the rock broke and moved.

Biological features

The dunes behind Bushy Point beach are dominated by beach grass, much of it planted by Project Oceanology students to help stabilize the dunes. Salt-spray rose and beach pea form scattered thickets on the back side of the dune. These plants are critical for stabilizing the dunes, which prevent waves from entering the marshes behind the dunes. Because of the sifting nature of sand, perennial plants are often uprooted, especially on the upper reaches of the beach. Thus many beach plants are annuals whose seeds move around during the winter to come up in a new location the next spring. They also are plants that can withstand the salt spray received from the high waves of storms. These include Russia thistle, seaside goldenrod, sea-rocket, and seaside spurge. Sand is a difficult place for plants to grow and walking on the plants is likely to kill them. Please walk on the dunes only where there are boardwalks. The vegetation at the beach on Mumford Cove has been severely damaged by mountain bicycles. If you ride a bicycle, please do not ride on the dunes.

Behind the Bushy Point beach and extending along the trail to the beach is a large tidal wetland. Marsh makes up most of the wetland, providing a highly productive food, shelter and breeding habitat for numerous invertebrates, fish and birds. Near the mouth of the Poquonock River, a saltwater intertidal flat provides a spawning and nursery area for winter flounder, as well as other finfish and shellfish.

Connecticut Minerals 8

Photo: Greg McHone.

Harold Stearns had $10 when he came to Wesleyan University in 1917, but quickly turned his hobby of rock collecting into a regular source of income. On weekends, he collected mineral samples from active quarries in the Middletown area and sold them to dealers. He described this project in an article published in 1980 in Lapidary Journal.

"The only way the quarries could be reached from the University at Middletown at that time was by streetcar across the Connecticut River to Portland and then on foot to the quarries," Stearns reported years later. "I did this with a knapsack on my back to carry specimens. On return trips I invariably cut the time necessary to catch the return trolley to a minimum and had to run with 25 pounds of minerals on my back."

Stearns came across jewel-quality specimens, one of which he was able to sell to the Boston Society of Natural History for $25. The value of that sum then is put in perspective by the cost of the boat trip he took on the *SS Richard Peck* down the Connecticut River to New York City. That trip, including a stateroom, cost one dollar—and led to more. The trip won Stearns a contract to collect minerals for the American Museum of Natural History and another $25 per week.

Stearns went on to become a very successful geologist as well, according to the biography by William Holden of Wesleyan University, the source for the preceding story. Not everyone who enjoys looking for rocks goes into geology, but most enjoy getting outdoors to places where rocks and minerals can be found. These are often also places where you can hike, observe plants and wildlife or enjoy the view—the types of activities people of all ages and backgrounds enjoy, not just "rock hounds."

Connecticut has a wide variety of interesting exposures of rocks and minerals that families can visit, including colorful gemstones, economic minerals once mined for use in building and manufacturing and, of course, rocks with their own stories to tell of our geological history. Mineral collecting is a bigger hobby than you might think. Several amateur enthusiasts' clubs meet regularly in Connecticut to discuss finds, trade rock and minerals or listen to expert speakers discuss various aspects of the hobby.

Learning about minerals can also be rewarding. Most people simply enjoy looking at the natural beauty of many stones. Of course, some are lured by the chance of maybe one day finding something truly valuable, spurred on by stories like the one about the part-time collector from Connecticut who recently sold a handsome specimen of Fluorite for thousands of dollars. More than a few geology students have paid their way through college by hunting down gemstones or panning for gold in Alaska.

Most, however, find simply spending time collecting or

trading minerals to be the real reward. There are many reasons to visit Connecticut rock and mineral locations besides chances for profit. The aesthetic and educational value of examining the underpinnings of our state are not things to place a value on, but rather like climbing a hill to admire a waterfall, peering through a telescope at the Moon or spotting a rare bird through a pair of binoculars. The rocks and minerals beneath our feet and along hillsides and quarries are the essence and structure of the place we all call Home—the earth. We live comfortably thanks to the resources the earth provides. Life is nicely decorated by the natural beauty of things like minerals.

To a geologist, the word "mineral" has a specific, scientific definition. To truly be a mineral, a substance must be solid, made by nature (not people), and crystalline, with a definite, inorganic chemical composition. Scientific definitions, however, leave out many substances called "minerals" in the popular media as well as in industry. In the broader sense, everything that is a "mineral resource," oil, coal, natural gas, cut and crushed stone, is a mineral even if it isn't solid or inorganic or a rock rather than an individual mineral.

Remember, never use a nail hammer to go after minerals. They are not made for it and can shatter or chip. A small chisel is helpful. Be sure that you and others nearby are wearing safety glasses, and watch for falling rocks. Finally, if there is a road nearby, stay well away from traffic and do not leave rock pieces where a vehicle or mower may hit them. Never park where you might block a lane or driveway. If traffic is too noisy, heavy, or fast, leave and come back on a quieter day.

Some common Connecticut minerals

The State Mineral, garnet, is quite common in metamorphic rocks of Connecticut. However, in many places garnet occurs as small and sparse crystals that can be hard to spot. Look for round, purplish-red crystals in relatively coarse-grained rocks.

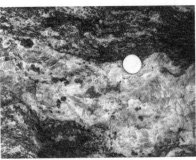

Burgundy Garnet in schist
Photo: Greg McHone

Garnet in pegmatite vein
Photo: Greg McHone

Muscovite is an abundant white mica in our rocks, but it only grows large in pegmatites such as at the White Rocks Quarry. Albite (white plagioclase feldspar) and orthoclase (pink to beige potassium feldspar) are found in many rocks, but they are most often collected from pegmatites and granitic rocks such as the Stony Creek Granite in Branford. Along with various feldspars, quartz is the most abundant rock-forming mineral, but only rarely will you find good crystal outlines. This is because quartz forms late in most rocks, so it has to conform to the shapes of the minerals already present.

Mixed with the above minerals in veins of pegmatite are several colors of tourmaline and beryl, which are most commonly black

Green Beryl in pegmatite
Photo: Greg McHone

Elbaite in pegmatite
Photo: Greg McHone

Tourmaline in Quartz
Photo: Greg McHone

and yellowish-green, respectively. When they appear in other colors such as blue-green, deep green, and red, they can be called gemstones. Good colors in combination with clarity and large size are what make them valuable, and sure to quicken the pulse of any mineral collector.

Pegmatite quarries of the Middletown District

The last active pegmatite quarry in the region was the White Rocks Pegmatite Quarry, which was worked by the Feldspar Corporation before it closed in the mid-1990s. Middletown has plans to develop the property into a new industrial zone. Hundreds of abandoned pegmatite quarries line the hills on either side of the Connecticut River between Glastonbury and Middle Haddam. Their concentration defines the "Middletown Pegmatite District," which provided a major industry for most of the previous two centuries.

Pegmatite is mainly composed of three minerals: colorless to gray quartz, pink to beige orthoclase (potassium feldspar), and off-white albite (sodium feldspar), along with minor volumes of gray mica, red garnet, black tourmaline, and numerous rare minerals. The abundance of the nearly-colorless quartz and off-white feldspar provides the very pale to white overall color of pegmatite, which is its most distinctive feature. In addition, the rock by definition has large mineral grains, usually the size of small pebbles but sometimes reaching dimensions of several inches to feet. Scattered pieces and pods of shiny mica often lend some glitter to the rock.

Pegmatites of the Middletown District were some of the first rocks in the world to be analyzed by radiometric methods to determine their absolute age, which is about 280 million years. Uranium-bearing minerals such as uraninite are used for this method. The age of the pegmatites is much younger than that of the metamorphic rocks around them, which are mostly older than 440 million years, but those rocks were partially recrystallized during the Alleghenian orogenic event that formed pegmatites.

Pegmatite is still quarried in North Carolina and elsewhere in North America for its feldspar, which is crushed and melted to form many useful ceramic products such as sinks, toilets, and spark plugs. Large mica flakes called "isinglass" were once collected from pegmatites for use as heat-proof windows in stoves and lanterns, and crushed mica is still used for heat and electrical insulation in electronic equipment. Where it can be concentrated in pure samples, the crystalline quartz in pegmatite can be used in making glass as well as silicon computer chips.

White Rocks Pegmatite Quarry

Precautions: The White Rocks Quarry is no longer active, and it has been partially reclaimed by filling in most of the pits. This has greatly reduced the hazard to visitors, but there are still walls with loose rocks that can fall, so you should not climb on any rock faces. The property is now owned by the City of Middletown, and it is being redeveloped for use by new light industries and possibly an electrical generating plant. Because the old quarry faces are along the side of a high hill, they will remain exposed and available for some time. However, mineral collectors should get permission at the Middletown City Hall before doing any digging.

Directions: The quarry area can be reached off Route 9 south by the Silver Street Exit. Turn left onto Silver Street. From Route 9 north, take the Bow Lane Exit, follow Eastern Drive to Silver Street, and turn right. Follow Silver Street eastward through the state hospital complex. About ½ mile past Silvermine Road, turn onto a gravel road to the right, which climbs the hill into the old quarry area (labeled Roadside Parks on the map). There are several old quarry faces along the road, which eventually ends in a loop.

What to See: The bright white color of the rocks in the quarry faces is due to the albite feldspar for which it was mined. You can also find abundant grayish quartz, beige orthoclase feldspar, white muscovite mica, and small patches of black biotite mica. Minor minerals include black tourmaline, red garnet, and several other minerals. Bring your mineral identification handbook to assist your examinations.

During the mining operations, the rock was blasted from the cliff faces, and large trucks would bring the rock pieces down the hill to a crushing mill. From there the crushed rock could be loaded onto railroad cars. The quarry company operated other mines in North Carolina, and the rock was brought to their mills for more crushing and separation of the different minerals.

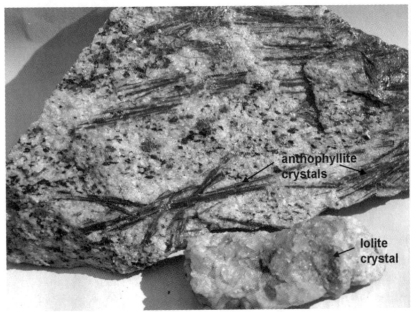

Photo: Greg McHone.

Iolite and anthophyllite in Haddam

Precautions: This locality is a low road cut on a side road with little traffic. Park so as not to block any vehicles that may come along, and stay off the pavement.

Directions: Take Route 9 to Exit 8 (Beaver Meadow Road), which is about 9 miles south of Middletown. Go eastward on Beaver Meadow Road just a few hundred feet or so and turn left onto Hubbard Road. Road cuts on the eastern side contain various minerals.

What to See: Monson Gneiss at this location also contains large pegmatite veins, in which are sprays of dark green anthophyllite (a metamorphic amphibole) as well as rare, deep blue gray iolite crystals (the gem form of cordierite). Mineral collectors have excavated pockets for the best iolite crystals, but if you pick up some of the loose rock pieces you might be rewarded with flashes of blue from this mineral. Iolite has the unusual property of dichroism, or two colors (blue and grayish brown), depending on from which angle it is viewed.

staurolite crystals

garnets →

Photo: Greg McHone.

Garnet and staurolite in Vernon

Some interesting and accessible outcrops for mineral collectors are present in the village of Rockville in Vernon, and they are conveniently reached from Exit 67 off I-84. This area is also the home of the famous Burgundy Quarry, which was also once a garnet mine. You need permission to enter that private quarry, but two road cuts are nearby, both of which are attractive because of their wide shoulders and out-of-the-way locations.

Precautions: Because these two localities are along public roads, you need to be especially careful about parking well off the pavement, and for watching over any dependents who are with you. Sundays are probably the quietest days for visiting, and you should avoid week-day commuter times.

Directions: Take Exit 67 from I-84 about 6 miles east of Hartford. Go to the staurolite-garnet locality by turning north on Route 31 (Bolton Road), then right (northeast) onto Route 30. After 0.6 miles turn right onto Industrial Park Drive West, then right onto Gerber Drive. Park just beyond the large road cuts on either side, near the road to the new ice cream factory. To get to the garnet locality, return back toward Exit 62 but pass under I-84. After 1.2 miles up the hill, turn right onto Dockerel Road and park so as not to block traffic. Walk back to the road cut on the north side of Route 31 and pay attention to vehicular traffic.

What to See: The Gerber Drive cut is in Bolton Schist, which is a local name for the Littleton Formation (named after Littleton, New Hampshire, the type locality). Abundant muscovite and biotite micas provide the glossy dark gray color, and small purplish red garnets are scattered like tiny beads throughout the rock. Staurolite is less abundant but larger, with elongated crystals up to an inch or more long, and with deep red brown, diamond-shaped cross sections. The Route 31 road cut is on Middletown Gneiss, and deep red garnets occur in layers toward the southeastern part of the cut. Some are large and clear, especially within quartz veins. Other minerals include tourmaline, quartz crystals, and apple green epidote, all of which attract collectors to this site.

Old Newgate Prison and Copper Mine

Precautions: This is a safe place for you and your family, but when you go down into the mine tunnel, watch your head under the low ceiling. Keep your children nearby and on the path.

Photo: Greg McHone.

Directions: Take Exit 40 off I-91, heading West on Route 20. Proceed about 8 miles and past the intersection of Route 187 with Route 20. Continue up the hill, and take a right at the traffic light. Head North on Newgate Road for about 1 mile to Old Newgate Prison historical park. Park just past the buildings on the left. The phone number of the museum is 860-653-3563, and hours are 10 a.m. to 4:30 p.m., Wednesday through Sunday, May through October. Admission is charged.

What to See: This old mine was worked during the eighteenth century for copper, which is in the form of copper sulfide minerals such as malachite and chalcopyrite. The ore minerals precipitated out of hot fluids that moved through the coarse sandstone during Mesozoic time. Although collecting is not allowed here, you can see some of the copper ore in the museum. Also, look for green copper minerals in some of the boulders of the rock wall next to the parking lot.

The prison that took over the mine has a colorful history as well, and it played an important Connecticut role during and after the American Revolution. Be sure to tour all the ruins.

Connecticut Museum of Mining and Mineral Science

Directions: Follow Route 7 north from Danbury to Kent, and go 1 mile north of the center of town to the museum entrance on the left side, off Route 7 and just past the Sloane-Stanley Museum (near Kent Furnace on the map). The mining museum is part of the Connecticut Antique Machinery Association. No admission fee, but donations are appreciated. Open Wednesdays through Sundays 10 a.m. to 4 p.m., May through October.

What to See: This museum has a fine collection of local rocks and minerals, and a darkened room that simulates the inside of a working mine tunnel and some old mining equipment. Special to this museum is a collection of bricks made in Connecticut, each identified by its style and origin. Brick making was once a big industry in Connecticut, but it has dwindled to a few companies.

You should also walk south and down the nearby hill below the Sloane-Stanley Museum (which has a wonderful collection of old tools) to see the old Kent Furnace, which for many years operated successfully to produce "pig iron" that was made into various tools and machinery parts. Iron ore was mined nearby from several bogs, where it occurs as layers of rusty minerals just below the surface.

Kent Falls State Park is just a few miles north along Route 7. This is a popular picnic place, and the waterfall flows over rock faces that were carved by the glaciers before 14,000 years ago.

Photos: Greg McHone.

Yale Peabody Museum of Natural History

Directions: In New Haven, Connecticut, take Exit 3 off I-91 (either north or southbound) onto the Trumbull Street Connector, and make a right turn at the second intersection onto Whitney Avenue (follow the posted signs to the Peabody Museum). The museum is located at 170 Whitney Avenue, at the corner of Whitney Avenue and Sachem Street, one block north of the intersection of Whitney Avenue and Trumbull Street.

What to See: The rocks and minerals exhibits at the Yale Peabody Museum offer a look at the geology of Connecticut—from bedrock to minerals—in one room. The room is dedicated to Benjamin Silliman, who added greatly to the Yale mineral collection in the early 1800s.

Exhibits describe the "big picture," the geologic history of Connecticut down to details of minerals from many localities. Start by entering the room on the right of the entry. Ahead are displays that explain how geologists define rocks and minerals. Minerals are inorganic chemical compounds with their own, distinct crystalline structure and are identified by their chemical composition, crystals and molecular arrangements. A rock, on the other hand, is composed of one or more minerals.

Keeping these distinctions in mind, cross the room to review the "big picture" of the geology of Connecticut. The large display describes the geologic history of the state. Maps on the wall above the glass case show topography, bedrock and tectonic history. The maps represent much of what has been learned of the state's geology by geologists from Benjamin Silliman to John Rodgers. Below the maps is a cross section that depicts the earth's crust in Connecticut. It shows the bedrock from west to east and the ages when layers of bedrock formed. On the floor below are actual specimens of bedrock, including metamorphic rocks from the northwest hills, examples of sedimentary and volcanic rocks of the Hartford Basin and the metamorphic rocks of eastern Connecticut.

Found among these rocks are many kinds of minerals. The glass cases to the right and left of the bedrock display include many of the minerals that formed in Connecticut over its long geologic history. To the right of the bedrock display is a case filled with beautiful forms of pegmatites, igneous rocks characterized by large crystals. Pegmatites are products of metamorphism. Many in the case are from Branchville, a small southwest Connecticut locality where a great variety of minerals have been found, including Albite, Beryl, Columbite, Microcline, Muscovite and Rose Quartz. Other Connecticut pegmatites on display include Chrysoberyl, Lepidomelane and Quartz from Haddam. There is Ilmenite from Litchfield, Cordierite from Plymouth, Samarskite from South Glastonbury, and Stilbite from Thomaston.

To the left of the bedrock display are three more cases that also include Connecticut minerals. The first is a display of limestones and marbles, formed from minerals deposited on ancient seafloors in the shells of early marine animals. There are several from the northwest hills, including Diopside and Tremolite from Canaan.

The next case displays minerals associated with traprock. In it are specimens of Anhydrite, Aragonite and Chalcocite from Meriden and Amethyst and Calcite from East Haven. Copper ore is also associated with basalt, and the museum has a 192-pound mass of copper from East Haven near the entryway.

The third case displays hydrothermal minerals. Most were formed with heat from ground water from deep in the earth's crust. A spectacular specimen of Barite from Cheshire can be seen in the entryway case, labeled Number 10.

Things to discuss: How do the appearances of Connecticut rocks reflect the geologic history of the state? How is it that such a great variety of minerals can be found in a small state such as Connecticut? Why might Benjamin Silliman have collected minerals as a professor at Yale College?

Responsibility & Resources 9

Photo: *Greg McHone.*

What do you need for your visits to geological locations in Connecticut? A geologist who is doing a scientific study in the field might carry a pack to hold maps of rock formations, roads, and topography, as well as a rock hammer and chisel, geological compass, magnifying lens, bottle of water, sample bags, field notebook, marking pens, an identification handbook, camera, cell phone, flashlight, small first aid kit, and often a Global Positioning System (GPS) receiver for recording precise locations.

Before heading out, a geologist will spend time reading about the regional geology and will study maps of the area. Field guidebooks are often available with descriptions of the best outcrops for rocks, minerals, and fossils that illustrate local geological history. Obviously, good hiking boots and other field clothes are needed, as well as a hat and bug repellent. And, as a matter of routine, he or she will always let someone know where and when they will be out "in the field." You don't have to be a geologist to pack similarly or follow the same routines.

Be sure to bring some maps with useful scales, starting with a road map, a trail map or park map if you walk very far, and perhaps a topographical map to show details of the terrain. Also, do not rely on your memory for descriptions and locations that you found in a field guidebook. Bring the book and other publications with you in the car, if not in your pack. A compass is not actually necessary for most excursions, but they are fairly inexpensive and very portable. Geologists use a special compass with a mirror for measuring ("sighting along") specific rock and terrane features, which are recorded as an azimuth or angle relative to true north.

A geology compass also has an extra dial with a pointer called an inclinometer, which shows the angle of dip, or tilt downward from level. You can get a similar compass from a good "outfitter" store, but just knowing which way is north may be all you need for finding your way on a map or country road. A GPS receiver can be very useful for its portable road map feature and for making records of places of special interest. For people interested in the new sport of geocaching, GPS receivers can be used to mark "virtual caches," places of geological interest to explore.

Environmental responsibility

Connecticut is filled with wonderful opportunities to explore natural history. These opportunities come with a responsibility to protect the state's natural resources and to preserve them so that others can explore and enjoy them as well. Learning a few basic "rules of the road" is a good way to conserve nature and to teach young people to enjoy the outdoors.

Collecting

Amateur geologists might wish to collect some rock samples to study or keep for a collection. Take photos instead. The simple fact is that collecting with photographs is an easier and more enjoyable way to experience the geology of Connecticut. Since digital photography has become more accessible and more affordable, it has become a preferred method of collection even among geologists. Taking photos is a very "low impact" way of exploring geology without degrading a resource or causing harm to sensitive natural areas. Many fishermen today, for example, choose to fish according to "catch and release" practices. Rather than removing it from a lake or river and killing it, many fishermen simply hold their catch up just long enough to get a picture, and then carefully return it to the water so some other fishermen can enjoy catching it on some other day.

Many geologists and amateur rock hounds choose to practice their own version of "catch and release" geology today. When you find an interesting geological formation or rock or mineral, leave it and take a picture away with you instead. You'll be surprised, but collecting pictures in a scrapbook or digital images on your computer can actually be a much better way to build a collection, and one more easily shared with friends and family. It's hard to lug a chunk of granite to school or to mail it to a friend. A scan of a photo or a digital image can be sent to people around the world, from relatives to professional geologists, in seconds.

Photos are great when accompanied with some written notes. Start and keep a log of your trips. Geologists seldom go anywhere without making notes of locations and conditions, and most amateurs find keeping a similar sort of log adds to the enjoyment of hiking in Connecticut to explore local geology. Many find that building a collection of photos and written notes is not just easier, but a more rewarding way to pursue their interest in rocks—much better certainly than keeping boxes of samples hidden away or perhaps left to be forgotten in a box in the closet or garage.

It is recommended that you never collect rocks or minerals unless you have an educational or professional use for them. Fossils should never be touched. Never attempt to make casts of fossil footprints; it can ruin them. If a fossil is mishandled, which can happen simply by not making an accurate record of some fairly arcane details, it immediately loses its scientific value. As much as we know about the natural history of Connecticut, there is always more that can be learned and nothing is more important than evidence garnered from fossils in their original condition.

If you must take a rock sample, please be considerate of others. There are many who will want to visit these same outcrops in the future. The safest way to collect a sample of a rock outcrop is to find a loose piece on the ground, where nature has already done the work of detaching it, and left a sample that can simply be picked up.

Basic safety

Never stop along highways or interstates to look at rocks. It is illegal in Connecticut without a permit and extremely dangerous under any circumstances. Whenever you take a trip to look at rocks, plan instead to visit sites in public areas such as state parks, where it is safe. Keep an eye on young children and pay attention to where you're going. Even a momentary distraction can easily lead to a twisted ankle or other accident. When hiking near cliffs or along steep rock formations, keep an eye out for falling rocks.

If you go to a location where there is a road nearby, be careful to stay off the pavement and do not leave rock pieces where a vehicle or mowing machine might hit them. Never park where you might block a lane or driveway. If traffic is too noisy, heavy, or fast, leave and come back on a quieter day.

It's always a good idea to pack water, a few snack bars, a first aid kit and a flashlight. If you have a cell phone, bring it along. And whenever you plan a hike in Connecticut, keep an eye out for plants like poison ivy. Looking for rocks often means bending over and putting hands and feet in places where its easy to come in contact with poison ivy. Often times you won't know it until the next day, when you start to break out. Ticks bites are also hazardous. Bring along bottles of poison ivy blocking lotion and insect repellent and wear long pants (as long as you can stand them) instead of shorts.

Another tip is to always have a backup plan ready. A little forethought and a good backup plan can often save the day when something unexpected comes up. If you find a location is too near a road or the terrain is too challenging or not appropriate for kids, have an alternate destination in mind. Jump back in the car and go on to "Plan B," and you won't end up in a position where you feel like you have to go ahead with a hike or an excursion that doesn't seem to be safe for your group or family.

Never under any circumstances take a nail hammer with you in the field or use one to strike a rock! Nail hammers are hard, but also brittle. They are made for relatively soft, steel nails. Whack a rock with one and a piece of a nail hammer could easily break off and fly up into your face or hit someone standing next to you. Rock hammers can be found at local nature stores. A glove on the hand that holds the chisel will reduce damage from the hammer and rock chips. Always wear safety glasses and do not allow people without glasses to stand nearby. Rock chips seem to know where to find your most vulnerable body parts. You can expect to use one or more Band-Aids from your first aid kit when collecting rock samples.

Private property & state parks

Remember: you must always have permission before taking a sample from private property or from any state lands or state parks. Make sure you have any required permits or permissions first before picking up a rock. Many private landowners cooperate with hikers and fishermen and amateur rock hounds, but this cooperation can be lost if mistreated. Stay off posted property. Be careful to park only in designated areas, not to block small roads and drives and never leave litter behind.

Managing environmental impacts

Whenever you go on a hike, in a state forest or park or private sanctuary, keep a few basics in mind to minimize the impact you have on sensitive environments. Stick to trails. Don't strike out across terrain that may be vital to wildflowers or other plants and animals. Long periods of time may be required for these areas to recover from being trod on by heavy boots and lugged soles or even bare feet.

Be careful around sand dunes, wetlands and marshes along the coast, and swamps, bogs and wetlands inland. Wetlands are highly productive natural systems, vital to sustaining nature's biodiversity. They are also fragile.

Don't carry anything in you don't plan to carry out. Better yet, pick up the litter that inevitably finds its way into even remote places. Don't expect trash cans to be provided; in many cases they are not. Keep a box of garbage bags in your vehicle instead and pack out whatever you find—empty bottles, discarded sandwich bags, containers or paper—anything that doesn't belong. Don't leave anything behind.

Resources

There are many publications about geology and Connecticut. Local libraries almost always have some Connecticut and United States government maps and bulletins produced over the past century or so. Even the oldest descriptions are usually still useful—after all, rocks took a long time to form and are slow to change! Some were written by geologists for other geologists and can be highly technical and filled with jargon. If you have some earth science in your background, they can also be very informative and surprisingly engaging.

Most bookstores sell field guides or handbooks for identifying rocks, minerals, and fossils. On a bookshelf they may be mixed among similar format books for identifying birds, trees, etc., in the northeastern or eastern states. A short list is provided below, but new examples come out frequently that may or may not be superior to the older guides.

Beware, books with the most colorful and spectacular photos of minerals rarely depict examples you will see in the real world, outside of museums. Common minerals (even the most interesting and valuable ones) are usually small, irregularly formed, and dirty. Consider buying a pocket guide to rocks and minerals that you can take with you on trips, and you might practice with it on some samples found around your home.

Books and web sites about Connecticut or regional geology are usually worth reading or browsing, including some good ones listed below. These may not be at levels amateurs can easily comprehend, but you will probably find some good illustrations of key earth science concepts and features that will be encountered on local trips. Discuss them with your family members and friends before you leave.

The Environmental and Geographic Information Center

The Connecticut Department of Environmental Protection (DEP) does much more than manage our state parks, clean up oil spills, and enforce environmental regulations (although its work in these areas is vital). The DEP also studies and manages wildlife and natural resources in our state, not only for the use of hunters and fishermen, but also to protect rare and endangered species. Within the DEP are the State Geological and Natural History Survey of Connecticut and the Environmental and Geographic Information Center (EGIC). The EGIC collects and disseminates information about our geology and natural resources published by the Geological and Natural History Survey and other sources.

The EGIC operates a little known, but remarkable place known as the DEP bookstore. The bookstore sells a wide variety of publications about Connecticut—a breadth of information available from few other states—and is a fascinating place to visit. You won't leave empty-handed. The DEP bookstore, located in the basement of the DEP offices at 79 Elm Street, Hartford, near the capitol, is open to the public. Parking is available on the street. A pass is required to enter the building and available at the security desk in the elevator lobby.

The bookstore has a large selection of maps, as well as books and bulletins on geology, paleontology, ecology, wildlife, boating, hiking, hunting, fishing, environmental protection, and other subjects appropriate for their mission. Some DEP publications may also be ordered online or through the store catalog.

EGIC has produced a series of CD-ROMs that can be used to make your own maps of a wide variety of geological and geographical information.

Maps and field guides

The Connecticut 2002-2003 Tourism Map is the official state highway map, published by the Department of Transportation. Free from state agencies and many information centers.

The Connecticut/Rhode Island Atlas & Gazetteer by Delorme is provided in a convenient scale (about one map inch per road mile) and format that includes most of our roads, waterways, and parks, with shading and some contour lines to represent hills and valleys. This atlas is available at many local bookstores or online.

The Atlas of Connecticut Topographic Maps is an 11 by 17-inch spiral-bound, black and white book containing all 116 USGS Connecticut topographic maps at 70 percent their original size, and is available at the DEP bookstore. Individual, full-sized *U. S. Geological Survey Topographic Maps of Connecticut* are also available at the DEP store.

The Generalized Bedrock Geologic Map of Connecticut is a full-color 8 ½ by 11 inch map of the rock formations of our state. On the reverse side is a brief geologic history of Connecticut, along with a list of other available geologic information. The map is available at the DEP bookstore for under a buck.

The Northeastern Region Geological Highway Map was published by the American Association of Petroleum Geologists in 1995. This colorful map presents rock outcrop information along the major roads of Maine, Vermont, New Hampshire, Massachusetts, Connecticut, Rhode Island, New York, New Jersey, and Pennsylvania. It is available online from the AAPG bookstore.

New England Intercollegiate Geological Conference (NEIGC) field guidebooks are published annually for fall field trip meetings. Each meeting is held in a different region of New England, but NEIGC guidebooks with trips in Connecticut are generally available from the Connecticut DEP bookstore, or from larger public and college libraries. NEIGC meetings with trips in Connecticut were held in 1968, 1975, 1978, 1982, 1992, 1998, and 2003.

Other groups such as the Geological Society of America and the National Association of Geoscience Teachers have published field trip guides that include Connecticut in 1991, 1993, and 1995. The field guide papers are often rather technical, but they include detailed maps with directions, as well as discussions and descriptions of many significant geological features. NEIGC field meetings are open to anyone, and information is available online.

Roadside Geology of Connecticut and Rhode Island, by James W. Skehan, Mountain Press, 2004. Part of a series of roadside geology books that already includes Massachusetts, New York, Maine, and Vermont/New Hampshire.

Rockhounding in Eastern New York State and Nearby New England, by Dan and Carolyn Zabriskie, published in 1993 by Many Facets, Inc. Mostly out-of-state locations, but easy to use.

Day Trips through Connecticut by Stan Gaby, printed by the author in 1979. A somewhat rough photocopied booklet available in a few specialty stores, such as Nature's Art in Salem. A little out of date now, but it still contains a lot of information about mineral collecting locations as well as places to bring your family on day trips. Sized for a field coat pocket.

AGI Data Sheets for Geology on the Field, Laboratory, and Office, by the American Geological Institute. A handy pocket-sized reference book with all kinds of geological information, conversion tables, classification charts, map symbol explanations, and other useful materials.

A Field Guide to Rocks and Minerals by Frederick H. Pough, sponsored by the National Audubon Society as part of the Peterson Field Guide Series, and published by Houghton Mifflin. Roger Tory Peterson was a great "birder" and Connecticut resident. This book provides such good mineral information that it is sometimes used for academic courses in mineralogy. Sized for a field coat pocket.

Simon and Schuster's Guide to Rocks and Minerals, sponsored by the American Museum of Natural History and published by Simon and Schuster. An easy-to-use field guide to many hundreds of rocks and minerals, many of which you will encounter in Connecticut. Excellent photographs and explanations. Sized for a field coat pocket.

Simon and Schuster's Guide to Gems and Precious Stones, edited by Kennie Lyman and published by Simon and Schuster. A fairly comprehensive description of gemstones, with pictures of the natural minerals as well as cut and polished specimens. Sized for a field coat pocket.

Books on geology and Connecticut

Lost Creatures of the Earth by Jon Erickson, published in 2001 by Checkmark Books. An account of the progression of life and the natural catastrophes that transformed it.

The Connecticut Valley in the Age of Dinosaurs: A Guide to the Geologic Literature, 1681-1995 by Nicholas McDonald, published in 1996 as Bulletin 116 of the State Geological and Natural History Survey of Connecticut.

Dinosaurs, Dunes, and Drifting Continents, by Richard Little, 3rd Edition, published in 2003 by Earth View LLC.

The Last Billion Years: A Geological History of the Maritime Provinces of Canada, by the Atlantic Geological Society, published in 2001 by Nimbus Publishing.

Written in Stone: A Geological History of the Northeastern United States, by Chet and Maureen Raymo, 2nd Edition, published in 2001 by Black Dome Press.

Lord Kelvin and the Age of the Earth, by Joe D. Burchfield, published in 1990 by the University of Chicago Press. An interesting account of political pressures in geoscience history.

The Man who Found Time by Jack Repcheck, published by Perseus Press in 2003. The story of James Hutton, whose revolutionary observations became the basis for modern geology.

The Map That Changed the World: William Smith & the Birth of Modern Geology, by Simon Winchester, published in 2001 by Harper Collins. How could we understand the earth without geological maps? William Smith overcame many challenges to make the first one.

Connecticut in the Mesozoic World, by J. Gregory McHone, 2004. A bulletin of the State Geological and Natural History Survey of Connecticut. There is a lot more to Mesozoic geology than just dinosaurs.

Stone by Stone: The Magnificent History of New England's Stone Walls by Robert M. Thorson, published in 2002 by Walker and Co. Professor Thorson of the University of Connecticut is an expert on geomorphology, the geological basis of our landscape.

A Guide to New England's Landscape, by Neil Jorgensen, published in 1977 by the Pequot Press.

A Sierra Club Naturalist's Guide: Southern New England, by Neil Jorgensen, published in 1978 by Sierra Club Books. Much good information about our plants and animals, as well as geology and the landscape.

The Smithsonian Guides to Natural America: Southern New England, published in 1996 by Smithsonian Books and Random House. Mostly ecology content, but with beautiful color photos of some Connecticut geological locations.

The Face of Connecticut: People, Geology, and the Land, by Michael Bell, published in 1985 as Bulletin 110 of the State Geological and Natural History Survey of Connecticut. Now out of print although not hard to find used. Some bookstores may still have copies.

Geologic History of Long Island Sound, by Ralph Lewis, from *The Tidal Marshes of Long Island Sound*, Bulletin No. 34, Connecticut College Arboretum.

Connecticut earth science museums
In addition to these museums, there are many small exhibits of Connecticut rocks, minerals, fossils, and geology in local schools and nature centers. Call the ones near you.

Science Center of Connecticut
950 Trout Brook Drive, West Hartford, CT 06119
860-231-2824.
Open daily until 5 p.m.
Admission is charged.

Dinosaur State Park
400 West Street, Rocky Hill, Connecticut USA 06067-3506
860-529-8423.
Open Tuesdays through Sundays, 9 a.m. to 4:30 p.m.
Admission is charged.

Peabody Museum of Natural History at Yale University
170 Whitney Ave., New Haven, CT 06520-8118
203-432-5050.
Open daily.
Admission is charged.

Connecticut Mining and Minerals Museum
Route 7, Kent, CT
Open Wednesday through Sunday, May through October.
Admission is free and donations are welcomed.

Old Newgate Prison & Copper Mine Historical Museum
115 Newgate Road, East Granby, CT
860-653-3563.
Open May through October.
Admission is charged.

Wesleyan University Museum of Natural History
Uxley Science Ctr, 4th flr, 265 Church St., Middletown, CT 06459
860-685-2244.
Open weekdays. Admission is free.

Connecticut State Museum of Natural History
University of Connecticut, 2019 Hillside Road, Unit 1023,
Storrs, CT 06269-1023
860-486-4460.
Open Monday through Friday, Sunday.
Admission is free and donations are welcomed.

Mashantucket Pequot Museum and Research Center
110 Pequot Trail, Mashantucket, CT 06338-3180
800-411-9671.
Open daily.
Admission is charged.

Connecticut rock and mineral clubs

Bridgeport Mineralogical Society
PO Box 686, Fairfield, CT 06430-0686

Bristol Gem & Mineral Club
PO Box 71, Bristol, CT 06011-0071

Connecticut Valley Mineral Club
14 Allen Ave., Westfield, MA 01085-2753

Danbury Mineralogical Society
PO Box 2642, Danbury, CT 06810-9998

Lapidary and Mineral Society of Central Connecticut
PO Box 476, Meriden, CT 06450-0476

Manchester Gem & Mineral Society
97 Moreland Ave, Newington, CT 06111-3525

New Haven Mineral Club
211 High Top Circle W., Hampden, CT 06514

Stamford Mineralogical Society
P.O. Box 4148, Stamford, CT 06907

Thames Valley Rockhounds
PO Box 2121, Salem, CT 06420